A Nebraska Bird-Finding Guide

A Nebraska Bird-Finding Guide

Paul A. Johnsgard

School of Biological Sciences
University of Nebraska–Lincoln

Zea E-Books
Lincoln, Nebraska

ISBN 978-1-60962-011-0   PAPERBACK
ISBN 978-1-60962-012-7   EBOOK

Zea E-Books are published by the University of Nebraska–Lincoln Libraries.

N

Contents

Author's Note

This publication had its origin as a self-published booklet in 2005. It was later adapted for on-line use by the Nebraska Bird Partnership (www.nebraskabirdingtrails.com). After producing 11 hard-copy printings, the materials were turned over in 2009 to the University of Nebraska–Lincoln's Digital Commons (http://digitalcommons.unl.edu/biosciornithology/51). This updated hard-copy version is its most recent incarnation. I greatly appreciate the efforts of Paul Royster and Timothy F. Jackson in bringing it about, and the help from people such as Linda Brown, Rachel Simpson and Eric Volden in the preparation of its various earlier versions.

Paul A. Johnsgard
April 12, 2011

Introduction

Birdwatching Throughout the Year in Nebraska

Persons living in Nebraska often feel that they are living in a cultural wasteland; its citizenry preoccupied with violent sports such as hunting and football. Yet many are unaware that they are actually residing in one of the prime locations in the entire world for observing and enjoying some of the most aesthetically appealing of all the world's biological attractions. The area around Kingsley Dam and Lake McConaughy, for example, is known to have attracted more than 330 bird species, including 104 breeders (plus 17 probable breeders) making it the third-most species-rich bird location in the interior U.S.A. (after Laguna Atascosa National Wildlife Refuge in southern Texas and Cheyenne Bottoms WMA in central Kansas). More impressively, the spring congregations of cranes and waterfowl along the Platte Valley have recently been ranked by Roger Pasquier (writing in *Forbes Magazine*, 1997) as the greatest bird spectacle on earth.

It has been estimated that bird-watching activities in the U.S.A. increased 155 percent during the 1990s, or at a more rapid rate than all other outdoor sports including walking, skiing and hiking, whereas fishing, hunting and tennis have all actually declined in popularity. Moneys now spent on wildlife recreation (over $100 billion) exceed total cash receipts from all livestock sales ("Wildlife Recreation," in *The Main Street Economist*, April, 2004). In 2001, 46 million birdwatchers spent some $32 billion. At least 63 million people in the U.S.A. feed or watch birds at home. In Nebraska an estimated 23.1 million dollars per year are spent on non-consumptive bird-related activities, and about 800 people have related jobs (*Bird Conservation*, spring, 1997, pp 6-8).

Every month of the year has its own bird-related attractions in Nebraska, as the following monthly breakdown will suggest. See also the migration calendar near the end of this book for more detailed information.

January

January in Nebraska is our coldest and dreariest month, and a good time for feeding birds and enjoying them through the windows. White-throated and white-crowned sparrows are welcome visitors to feeders now, as are American goldfinches and other typical finches, such as purple finches (rarely) and (increasingly) house finches. Dark-eyed juncos and eastern or spotted towhees scratch industriously in the snow for food, and in the countryside species such as horned larks, American tree sparrows, Harris' sparrows, and sometimes Lapland longspurs gather in open fields to search for seeds. During some years red-breasted nuthatches appear at suet feeders to join wintering downy, hairy, and even less often red crossbills and evening grosbeaks make their appearances, especially in western parts of the state. In the panhandle and Pine Ridge other winter visitors might include Steller's jays, Bohemian waxwings, Clark's nutcrackers, mountain chickadees, Cassin's finches and gray-crowned rosy finches.

Northern cardinals now begin to sense the lengthening days, and by mid-January a few males may begin to sing, to be joined later in the month by European starlings and sometimes a few precocial house finches.

For the adventurous birders, this month offers the temptation to drive to Kingsley Dam near Ogallala for the annual spectacle of several hundred bald eagles that annually gather there, and the chance to see some rare waterfowl, loons or gulls as well. This is the best time for seeing rough-legged hawks and other uncommon to rare northern visitors, such as snowy owls.

1

2

February

By February the days are perceptibly lengthening, and the end of winter seems almost in sight. Winter birds still gather at feeders, but American crows are on the move northward, as are vanguard individuals of American robins, red-winged blackbirds, and bluebirds.

By shortly after Valentine's day the first sandhill cranes can be expected to arrive on the Platte river, at least if it has become ice-free. They usually arrive on a south wind and with clearing skies, probably having flown from their wintering areas in Texas and New Mexico in a single day. Common mergansers begin to appear on the Platte too, as well as on larger reservoirs. Almost simultaneously common goldeneyes start to appear in these same locations.

As lakes and reservoirs slowly become ice-free bald eagles appear throughout the state's waterways. They feast on winter-killed fish that become available, often perching on ice blocks or on trees along the shoreline. Harlan County reservoir and Sutherland reservoir in south-central Nebraska are good places to look for eagles, as well as Lake McConaughy in the Panhandle and larger eastern reservoirs, such as Lewis and Clark Lake, Branched Oak Lake, and Cunningham Lake. DeSoto National Wildlife Refuge often has good numbers of bald eagles in February, and its visitor center allows for easy and comfortable eagle-watching, as does the J-2 powerplant near Lexington.

Great horned owls begin nesting in February, and should not be disturbed by playback of owl calls from this point onward through their nesting period.

March

March is perhaps the most exciting month of the year for Nebraska's birders. By the first week or two hundreds of thousands if not millions of geese (snow, Ross', Canada, cackling, and greater white-fronted) will have arrived in the eastern Rainwater Basin of southeastern Nebraska (centering on Clay County), and on a few basins farther west, such as Funk Lagoon. Unbelievable hordes of geese, as well as early duck migrants such as northern pintails and mallards crowd these marshes and the skies above, performing dizzying courtship flights and endless feeding flights to nearby fields for foraging. These great flocks of waterfowl usually peak by the middle of March.

During March the sandhill cranes begin to build up to a peak of about a half million birds in the central Platte Valley, spending the daylight hours feeding in cornfields and wet meadows, and roosting in shallow water around sandbars and islands in the wider portions of the river. Flights to and from their roosts occur at about sunset and sunrise, although cloudy skies may cause earlier evening flights and later morning departures, just as clear skies and a full moon may allow the birds to remain in fields for a longer period.

Mid-March is the target date for the Audubon Society's annual Rivers and Wildlife Conference at Kearney, drawing birders from around the country and the world. Finding space in one of the several crane observing blinds is difficult then, but crane viewing from the road or from viewing platforms near Alda, Gibbon, and a hike-bike bridge near Fort Kearney are all options for the less lucky individuals. All of these experiences are memorable, but watching cranes land nearby at a blind near an evening roost, or seeing them depart at sunrise, provides a thrill unmatched anywhere in the world.

As March draws toward a close the second wave of duck migrants arrive, including American wigeon, gadwall, wood ducks, green-winged teal, redhead, canvasback, ring-necked duck, bufflehead, and hooded merganser. The final wave, with blue-wing teal, northern shoveler, ruddy duck, and perhaps cinnamon teal, complete the roster. Other water birds also arrive in March, including American coot, pied-billed and eared grebes, and western grebes, plus American white pelicans and double-crested cormorants.

The resident red-tailed hawk population is now supplemented by migrant red-tails as the rough-legged hawks begin to head north, and northern harriers (males first) begin to appear in the state. This is a good time to watch male harriers perform their circular territorial display flights (from which their generic name Circus derives), and to start looking for nest-building behavior in red-tails. Ospreys begin to funnel through the state in locations where bald eagles have been prevalent, and prairie grouse (sharp-tailed grouse and greater prairie-chickens) begin to assemble at their traditional display grounds, or leks.

April

The first half of April marks the mass departure of sandhill cranes from the Platte Valley, the arrival of whooping cranes, and the peak of mating activities by the prairie grouse. Seeing the great flocks of sandhill cranes rise up from the river one clear morning, circling while calling excitedly, gradually gaining altitude, and finally disappearing from view even though their great trumpeting voices still drift down like a vast but unseen angelic chorus, is an experience of a lifetime.

Whooping cranes drift into the state in small family groups or flocks of up to about ten in size, stopping in the Platte Valley or in other wetland areas such as rivers in the Nebraska Sandhills or the Niobrara Valley. Rarely they may feed in the company of sandhill cranes, or roost among them at night.

The second week of April (often the third week in northern portions of Nebraska) is the best period for seeing prairie grouse display, for during this rather short time the majority of females visit the leks for mating. This sets off a frenzy of display activity and fighting among the males, to determine which will gain the opportunity to fertilize the suddenly available females. Somehow the females can determine the identity of the most virile and dominant male of each lek, and seek out this individual for their mating. Their leks are usually in areas of native grassland that are well away from tall trees, and often on short grass-covered hilltops. Blinds at locations such as at Halsey and McKelvie national forests, Crescent Lake, and Valentine and Fort Niobrara national wildlife refuges provide excellent sites for watching these activities.

The second half of April usually coincides with the peak of shorebird migration in Nebraska, as well as the arrival of the first insectivorous songbirds, such as the swallows, vireos, and warblers. This is an exciting period, as waves of plovers and sandpipers arrive at wetlands, and the skies overhead become clotted with Franklin's gulls, swallows of several species, and the blooming wild plums begin to reverberate with the songs of house wrens, rose-breasted grosbeaks, brown thrashers, and early-arriving warblers such as yellow-rumps.

May

May is simply magical in Nebraska. During the first two weeks of this month the peak of songbird migration occurs, with gorgeously plumaged warblers and plainer sparrows of infinite variety frustrating the observer by quickly scurrying about among tree canopies or skulking in the grass. Often one beautiful species will be present in a wooded habitat or prairie on one day, only to be replaced on the next day with a new and equally interesting one. Butterflies and early spring flowers begin to appear, and persons tending their bluebird nestboxes can expect to peer in one day and see the female huddled down on a brood of squirming youngsters. Broods of Canada geese start to materialize on farm ponds and in city park lagoons, and in city backyards house wrens are simultaneously singing and feeding new broods.

Early May marks the International Migratory Bird Day and the time of the annual Audubon Birdathon in Nebraska. Then participating birders may compete to produce the largest single-day bird list possible, with donors providing challenges by donating moneys to the local Audubon chapters. These funds aid prairie preservation and the purchase of instructional materials dealing with nature for distribution to grade schools.

June

June is the perfect time to be in the field in Nebraska; the long days allow for after-work birding, and this too is the month when Breeding Bird Surveys need to be carried out. People monitoring bluebird trails are busy then too; first or even second broods of bluebirds and tree swallows are likely to be out, and boxes need to be patrolled to prevent depredations by raccoons or invading house wrens or house sparrows.

By the first of June the most tardy of the spring migrants, the common nighthawk and the black-billed and yellow-billed cuckoos, will have arrived in Nebraska, and the last arctic-breeding shorebirds should have departed. (By the end of June the appearance of any such arctic shorebirds in the state may actually represent early fall migrants, namely those individuals that were unsuccessful breeders and are already heading back south.) Summer evenings will be enriched by the distinctive territorial calls of nighthawks, chuck-will-widows, whip-poor-wills, and (in western Nebraska) poor-wills. The cuckoos (often called "rain crows" in Nebraska) may call from their hidden locations as evening twilight or afternoon thunderstorms approach. Generally, however, singing by birds diminishes in June, as the birds become preoccupied with nesting, and only such multiple-nesting species as house wrens, or those males that lost their initial mates and must quickly acquire new ones, are likely to be singing at full strength.

July

July is too hot for most outside activities in Nebraska, and birding activities are best confined to early morning walks, when a few die-hard songsters such as house finches may still be active. Second broods of many species will now be appearing, and early fall migrants such as long-billed curlews and cliff swallows will be starting to gather for migration. It is amazing that the young of birds such as these can be ready to undertake flights of up to several thousand miles only a few weeks after hatching in the case of the swallows. Some multiple-brooding birds such as mourning doves and barn swallows will still be industriously fledging early broods and starting new ones soon thereafter, producing four or perhaps even five broods in a single season before running out of time. On the other hand trumpeter swan cygnets being reared in Sandhills marshes will only be about half-grown by July, and the approximate 100-day fledging period will require most of the summer before the cygnets are able to take their first flights.

By early July the brown-headed cowbird females, who may have already laid 40 or more eggs in the nests of unlucky hosts, will finally have become too exhausted for further laying. Thus, late-nesting sparrows and warblers may be spared the fate of having to raise a cowbird chick with their own young, which often results in the starvation of the host's chicks as a result of the cowbird's gluttonous appetite.

August

With the arrival of August the first sense that summer is almost over begins to take hold; the cooler mornings and the gathering flocks of swallows along telephone lines provide an early warning system that the good times are nearly over. This is a period when arctic-nesting shorebirds begin to filter into wetlands having muddy and sandy shorelines, and a chance for the birder to test his or her skills at identifying the maddeningly similar immature and fall-plumaged "peep" sandpipers, or the equally frustrating "confusing fall warblers." This is a perfect time for 10-power binoculars or spotting scopes and tripods, with their magnifications set at maximum power, and all the available field guides close at hand.

September

The blue skies of September bring not only occasional cumulus clouds to Nebraska, but also clouds of early fall migrant "blackbirds" (red-winged blackbirds as well as grackles, cowbirds, and starlings). As the trees begin to turn color the first frosts send insect-eaters such as warblers and vireos scurrying southward, and set the stage for the great migrations of the larger birds. Migratory hawk species, such as Swainson's hawks, some non-residential red-tailed hawks, as well as Mississippi kites and turkey vultures begin to assemble and ghost southwardly. A keen observer may scan the sky with binoculars for skeins of geese or ducks, or perhaps may train a spotting scope on the face of a full moon some evening, and see fleeting silhouettes of distant birds crossing in front of it.

This is a good time to wander aimlessly through the woods; mosquitoes and chiggers are no longer a problem, and the dying leaves begin to allow a better look into the upper levels of the forest canopy. Escaping into the country also permits one to avoid the screaming hordes of football fans that are attracted to stadiums like ants to honey, mostly wholly unaware that the greatest visual spectacle on earth is passing by overhead.

October

October is the most colorful month of the year in Nebraska; the peak of fall color occurs about the middle of the month, and many wonderful birds are moving through the state's wooded habitats at the same time. Not long ago the peak of the arctic goose migration occurred in October, as several million snow geese would funnel down the Missouri Valley, and Canada and greater white-fronted geese would appear in the central and western parts of the state. Recently, however, the fall migrations of these geese have peaked later, often during the first week of November, perhaps as a reflection of global warming trends.

Nevertheless, October brings with it a major movement of larger migrants, from hawks to the early duck migrants such as blue-winged teal and shovelers, and many of the more tardy shorebirds. Sandhill cranes begin to appear in marshes of the Nebraska Sandhills. Few of them use the Platte Valley, since intensive waterfowl hunting activities there make the area unsafe for cranes.

November

During November the birding season comes almost to a close; a few northern shrikes and rough-legged hawks are now arriving, and migratory sparrows such as American tree sparrows and Harris' sparrows start to materialize in shrubbery and thickets. Eagles start to invade the state in good numbers, tagging along with the flocks of ducks and geese, and occasionally snagging a sick or wounded one. On clear and calm evenings great flocks of geese can be heard overhead bound for unknown destinations using clues that we can only try to imagine.

Owls begin to set up their breeding territories now. Hardy birders will find that this is a good time for playing recordings of various owls after dark, then listening for responses. Barred owls respond especially well, often flying into a tree directly above the tape recorder.

December

December is in many ways the cruelest month; the days are so short that there is little time after work to go afield, and few birds to see when one does so. Yet it is a month for planning a Christmas Bird Count with close friends, and perhaps making out a Christmas list of bird-related gifts to present to friends and family, or perhaps hope to receive from them. The Christmas Bird Count, sponsored by the National Audubon Society, is the last organized birding activity of the year in Nebraska. Counts typically occur at about 10-12 sites, usually including Beaver Valley, Branched

Oak-Seward, Calamus-Loup, Crawford, DeSoto N.W.R., Grand Island, Lake McConaughy, Lincoln, Norfolk, Omaha, and Scottsbluff.

December is also the time to make out summaries of yearly bird lists, at least for people who keep such lists, and a time to start planning birdwatching trips for the following year. It is also not a bad time to consider a trip to Florida, southern Texas or even the tropics of Costa Rica, for a chance to get a flavor of how rich the bird life can be in places such as these. It is a time to look back on all the wonderful experiences of the previous year, such as that stunning scarlet tanager singing in the treetops, the spine-chilling sounds of sandhill cranes approaching their roosts, or a ruby-throated hummingbird that danced momentarily in the sunlight like a tiny sprite before it disappeared in the twinkling of an eye.

Fundamentals of Birdwatching

In recent decades birdwatching has become one of the major recreational activities engaged in by Americans, with over sixty million persons now participating in this combination of intellectual activity, exercise, sport, and, in its most academic form, science. Its attractions include the facts that it can be equally enjoyed by absolute novices or lifelong devotees, it can be practiced by people ranging in age from a few years to elder citizens, and it requires very little in terms of special equipment. Finally, it is essentially of inexhaustible interest, given the fact that there are over 9,000 species of living birds in the world, and nobody could ever live long enough to learn about or even observe more than a small proportion of them.

Many birdwatchers are determined to see as many species as possible, and develop "life lists" that may be subdivided into state lists, yearly lists, or other categories. Such persons tend to call themselves "birders," and for them "birding" often has a competitive aspect, in which a maximum number of species that can be detected (seen or heard) and identified in a single day or year has a particular attraction.

Others are content simply to enjoy the birds attracted to nest boxes, feeders or water sources set up in their back yards, thus reveling in the fascinating behaviors, diverse plumages, nesting activities, or other facets of bird life that may be visible from their patio deck or through their windows. Still others become avid bird-banders, engaging in a kind of avian lottery, and trying to recapture birds already banded by themselves or others. By banding and releasing birds they hope that, like placing a message in a bottle, somebody in the future may find the bird's carcass or perhaps recapture it, and by reporting it provide evidence of the bird's movements and longevity following its banding.

Finally, there are the professional ornithologists who, like other kinds of zoologists, might be interested in anatomy, genetics, physiology, classification, ecology, behavior, or any of the other branches of biological science, and tend to be educators or museum scientists. There are only a few thousand such persons, and perhaps a few hundred additional bird-related professionals who lead bird-watching tours all over the world. Still others may write about or photograph birds for a living, or perhaps draw, paint, sculpt, carve, or otherwise depict birds in some art-related field.

Optical Equipment and Acoustic Aids

Depending on one's area of interest, differing kinds of equipment and resource materials may be needed. The most popular type of bird appreciation, field observation and identification, requires only a minimum of equipment. Probably the most important of these are a pair of binoculars and an identification reference, or "field guide."

All binoculars are identified as to their magnification power and the diameter of their front ("objective") lenses; thus a pair of 8 x 40 binoculars has eight-power magnification and a front-lens diameter of 40 mm. Doubling the diameter of the objective lens quadruples the light transmission of the lens, but also increases its weight; the magnification of the binoculars does not influence their weight. Generally, to insure adequate light-gathering power under dim-light conditions, the objective lens should be at least five times greater than the magnification. This ratio (which would be 5.0 in 8 x 40 binoculars) is called the "exit pupil" index, since it is a measure of the diameter of the circle of light exiting the rear of the binocular and thus entering one's pupil. "Relative brightness" is calculated as the square of the exit pupil index. However, this index (the larger the better for maximum brightness under dim conditions) provides

only a general guide to actual image brightness, and unless one is using the glasses in dark forests it is not a major consideration, and the lenses' light transmission efficiency is also important.

Often of greater interest than the exit pupil index is the binocular's angle of view, or maximum visual field, which is usually indicated in degrees. Binoculars typically have fields of view ranging from five degrees (263 feet at 1,000 yards) to 8.5 degrees (446 feet at 1,000 yards). "Wide-angle" binoculars (with fields generally of seven degrees or more) are generally better choices, but often are heavier than standard binoculars because of their larger prisms. Wide-angle binoculars sometimes also have increased distortion at the edges of the field of view.

Choosing a suitable pair of binoculars requires special care. Not only because a substantial outlay of money is involved, but also the choice made plays a large role in determining how effectively one will be able to locate and identify birds. These may be moving rapidly, mostly hidden by foliage, or so far away that considerable magnification is needed to observe their critical identification features. The subject of binocular choice is thus a complex topic, dictated in part by one's budget, and in part by the kind of primary birdwatching that is contemplated (for example, in forests, along lakes, or under dim-light conditions). Price ranges of binoculars range from less than $50 to several thousand dollars; my own two favorites include one that I bought at a tag sale for $19, and another purchased new for $150. Several friends have pairs costing close to $1,000, and at times I envy them, but generally one can find a suitable pair for less than $200. The Bushnell Birder 7 x 35, at about $75, is acceptable for starting birders on a strict budget.

When handling a pair of binoculars a prospective buyer should test for double images (produced when the paired optics are out of perfect alignment), colored fringes around the surfaces of objects visible at the edges of the field (a sign of chromatic aberration), apparent curvature of straight-line objects such as rooftops or utility poles (a sign of optical defects), and distortion, fuzziness, or light flare at the edge of the field when looking at a bright point of light such as a star. The ease of turning the focusing wheel (especially in cold weather) may be important, but the adjustment hinge for maintaining proper inter-pupillary distance (to correspond with the width of one's eye spacing) should not be too loose. A comfortable grip and weight are also considerations, and the neck-strap should be sturdy and comfortable as well. Better quality binoculars have anti-reflection coatings on all internal surfaces, are water-resistant, and may be internally "purged" with gasses that prevent fogging in wet weather.

For eyeglass-wearers the eye-relief distance (the distance between the binoculars' rear lens and the farthest point behind the lens at which the binoculars' entire field is visible) is more important than the binoculars' actual field of view. An eye relief of about 20 mm. is needed for the eyeglass user to see the entire field and avoid "tunnel-vision" vignetting effects. This eye-relief measure is rarely indicated in the binoculars' specifications, but possible vignetting effects should always be personally checked by prospective buyers who wear eyeglasses. Long eye-relief binoculars are usually limited to lower-power models and generally lack wide-angle features, so there is a trade-off in making such selections.

My own preferences are either 8-power or 10-power binoculars; the higher power glasses tend to be heavier, have a reduced field of view, and are harder to hold steady, but are excellent for viewing distant objects. Seven-power binoculars with a large exit-pupil ratio of at least 6.0 may be favored by those people who usually birdwatch in dense forests and require maximum light transmission capabilities. Armored binoculars (with a rubber coating) are more resistant to water and physical shocks than regular ones, and may be easier to hold, but are about two ounces heavier than non-armored models. Waterproof models are probably worth the extra cost. Binoculars with roof-prisms are straight-tube in shape; those with regular prism design ("porro prisms") are usually wider at the front than in back, but occasionally in very small glasses are wider at the rear than in front. Roof-prism models are always more expensive, and often have narrower fields of view than porro-prism models. Yet, they average somewhat lighter and are less likely to go out of alignment, a condition that results in a double-image visual effect.

One should strictly avoid zoom binoculars having variable power; they are always heavy and tend to have terrible optics. One should also avoid binoculars with tilt-lever focusing; they usu-

ally don't focus very closely, and, being generally cheap, tend to be loosely and poorly constructed. Similarly, avoid "universal focus" or "focus-free" optics that are pre-focused for moderate distance; it is impossible to adjust focus for close or extremely distant objects. Similarly, birders should buy binoculars that focus at least as close as about 10 feet or so; some models now focus down to less than 5 feet. Other binocular types to be avoided are those with individual-focus (as opposed to central-focus), and those that are either too small or too large to be easily handheld and focused. Persons with small hands or children may, however, find small glasses more comfortable than standard size ones.

In a review of binoculars published in The Living Bird in their Autumn 1995 issue, the mid-priced binoculars ($200-500 or more) most favored were the Swarovski 8 x 30, and either the 8- or 10-power Swift Ultra Lites. The Bausch & Lomb Custom Compact (7 x 26) was judged the best of the small design models. For eyeglass wearers the only recommended models were some of the more expensive ones, such as the Nikon Talon and Wolverine models. High-end favorites included the top-of-the-line models by Zeiss (Dialyt), Leica (Ultra), Bausch & Lomb (Elite), and Swarovski (SLC). The Nikon 9 x 30 Execulite is much less expensive but ranks close to the top-level models, although it is not recommended for eyeglass-wearers. The Swift Audubon 8.5 x 44 is still less expensive and may be a suitable choice for those wanting near focusing, but it too has limited eye-relief. The Celestron Ultima series is even less costly and a very good choice for price-conscious people; the popular 7 x 44 model has 19 mm of eye-relief and thus is quite suitable for eyeglass wearers.

Many "hard-core" bird-watchers own at least two pairs of binoculars of varied designs and powers, and some also invest in "spotting scopes" rather than using very high-power binoculars. These scopes usually are of 20-power to 30-power, and when used with a sturdy tripod are wonderful for identification of birds at great distances. However, they are quite expensive; the better models usually costing $500-$1,000 or more, and weigh several pounds. Zoom optics on the best of these scopes can be excellent, but there is a reduction in light transmission at the higher magnifications. Various models of Kowa scopes (TSN-2, TSN-4) are especially favored by birders, but are fairly expensive. The Bushnell Spacemaster may be adequate for birders on a budget (Living Bird, spring, 1984), but the zoom eyepiece should be avoided. Some scope eyepieces are designed for eyeglass-wearers; Celestron calls theirs "long view" eyepieces and Kowa's are called "LER" eyepieces. A solid tripod, such as the 8-pound Bogen 3021 or the lighter Davis & Sandford RTS, is essential for use with spotting scopes.

Binoculars and other optics are often reviewed in the periodical "Better View Desired," published by Whole Life Systems, P.O. Box 162, Rehoboth, NM 87322; free sample issues are sent on request, and back issues are available. Binoculars and telescopes can be purchased in many sporting-goods and discount stores, but wise shoppers will investigate mail-order sources. B. & H. cameras of New York (213/807-7474) is one of the more trustworthy New York discount houses, but considerable care should be exercised when dealing with any of these outlets, which are notorious for their bait-and-switch tactics. An excellent catalog, with good prices and much technical information on choosing binoculars and spotting scopes, including comparative weight, field-of-view, eye-relief and near-focus data, is available from Eagle Optics, 716 S. Whitney Way, Madison, WI 53711 (608/271-4751). Their staff also provides excellent advice. The American Birding Association's (A.B.A.) sales catalog (800/578-0607) is quite informative, but their prices are not quite so competitive as are those of Eagle Optics or the New York discount houses. Other reliable mail-order houses that provide informative catalogs are Christophers, ltd., 2401 Tee Circle, Suite 106, Norman, OK 73069 (800/356-6603) and National Camera Exchange, 9300 Olsen Highway, Golden Valley, MN 55427 (800/624-8107).

Besides optical equipment, many bird-watchers use cassette tape-recorders or tape-players. The A.B.A. sales catalog (see phone number above) has a very good selection of birdsongs on cassette tapes and CD recordings. Pre-recorded tapes are extremely useful for learning birdsongs and other vocalizations, and may also be valuable tools for playback in the field, to stimulate responses from species that are normally difficult to see, such as rails or owls. Indeed, for such birds playback of vocalizations may be the only practical means of detecting the species' presence in an area. Various recorded "field guides" to bird songs are now commercially available, including "Eastern/Central Bird Songs" and "Western Bird Songs," designed to be used with the Peterson field guides. There is also a "Guide to Bird Songs" for use with the National Geographic Field Guide, all of which are available as cassettes or CDs. Several other audio guides exist, including a series called "Birding by Ear" that organize songs according to common traits.

There are also CD-ROMs that provide both vocalizations and illustrations, usually along with individual range maps and other information on each species. One of these is "Birds of North America" by Thayer Birding Software (for Windows only) that not only provides quizzes and side-by-side species comparisons, but also includes the entire text of The Birder's Handbook by Paul Erlich et al. (see below). Another, the "National Audubon Society Interactive CD-ROM Guide to North American Birds" also includes quizzes, and is available in either Windows or Macintosh versions.

Some CD-ROMs offer convenient regional checklists for personal record-keeping, or provide national, continental, or even world-wide checklists or distributional information in computerized form. The A.B.A. catalogs list many of these, as well as a variety of videos on field identification.

Reference Materials

Besides optical equipment, every person interested in birds needs some references for aid in identifying birds and, depending on level of interest, learning more about them than their names. At minimum, a field guide suitable for carrying along in a pocket or pouch is needed. Since the development of the first modern field guide by Roger T. Peterson in the1930s, a veritable host of field guides have been published. Only a few of these are outstanding, and indeed many persons can get by with a single guide.

Persons in Nebraska are faced with the fact that our state lies in the transition zone between North American eastern and western avifaunas, and as a result neither R. T. Peterson's eastern or western field guide is entirely adequate for this region. Based on over 40 years of teaching ornithology, I have come to recommend that beginning students use *Birds of North America*, by Chandler

Robbins and others, published by Golden Press. It is relatively inexpensive, easy to use, and covers all North American birds. Its paintings and range maps are adequate, and its organization is excellent. For more advanced birders I usually recommend the National Geographic Society's *Field Guide to the Birds of North America*. It is larger and considerably more expensive (about $25.00) than the Golden Press guide, but has better paintings and range maps, and shows a much larger number of plumage variations (races, plumage "morphs" such as melanistic or leucistic variants), and has better descriptive captions. However, it is somewhat daunting for beginning birders, and is not a good "starter" choice. A fairly new book by David Sibley is much more complete, much larger, and is more a home reference book than field guide. It is called *The Sibley Guide to Birds*, and is notable for its many views of diverse plumages and different viewing angles. It is published by the National Audubon Society and A. A. Knopf. His smaller and more recent eastern and western guides are much easier to carry in the field, and the western guide covers virtually every species likely to be seen in Nebraska. It appears to be an excellent choice for starting or advanced birders.

Field guides that should be avoided at all cost are those that use color photographs rather than paintings, and which are usually organized by predominant plumage color, such as the various Audubon guides. I always tell my students to throw away any such guides they may already have, as they are generally worse than useless. A guide, designed much like the Golden Press guide, was published in 1997 and titled *All the Birds of North America*, but has not proven very popular. A newer (2001) field guide by Ken Kaufman, *Birds of North America*, uses computer-enhanced photos, and seems a useful introductory guide. However, the two newest (2003) eastern and western guides, by David Sibley, the *Sibley Field Guide to Birds*, seem much the best of any now available. The one for western North America covers all of Nebraska's non-accidental species.

Bird-finding guides to many states are now becoming increasingly popular; many of these are described in a catalog sold by the American Birding Association (see address below). The classic if now badly outdated book in this area is O. S. Pettingill's volume *A Guide to Bird Finding West of the Mississippi* (Oxford University

Press). There are more than 400 Nebraska birding sites posted on the following website: http://nebraskabirdingtrails.com/, which was primarily written by the present author for the Nebraska Partnership for All-bird Conservation and duplicates much of the information here. The locations on the website do include some not present on this hard-copy version (mainly privately owned sites), and often have detailed site maps and tourism-related information such as nearby accommodations.

The Nebraska Game and Parks Commission has recently (2004) published a special 178-page issue of NEBRASKAland titled "Birding Nebraska," which details over 60 major birding sites in the state, and has a complete checklist of the state's birds. It earlier (1997) published a 96-page booklet by Joseph Krue, titled *NEBRASKAland Magazine Wildlife Viewing Guide*, which includes descriptions of 68 wildlife viewing sites in the state. It is out of print but may sometimes be obtained from used bookstores. The Game & Parks website also has much useful information for birders; including a photo gallery of Nebraska birds: http://outdoornebraska.ne.gov/. The Patuxent Wildlife Research Center has a valuable website, "Patuxent Bird Identification Info Center," with photos, songs, videos, maps and life history information on most North American birds: www.mbr-pwrc.usgs.gov/id/framlst/framlst.html. A corresponding and equally useful identification guide is available through Cornell University's Laboratory of Ornithology website: www.birds.cornell.edu. The Nebraksa Bird Library website www.nebraskabirdlibrary.org is developing a similar guide.

The 200-plus breeding species occurring in Nebraska were documented in my book *The Birds of the Great Plains: The Breeding Species and their Distribution* (University of Nebraska Press). A county-by-county historical summary of the breeding birds of the state may be found in James Ducey's *Nebraska Birds: Breeding Status and Distribution*. A comprehensive book, *Birds of Nebraska*, by Roger Sharpe and others, is now available (Sharpe et al., 2001). Breeding bird surveys done in Nebraska during the 1980s (1984-1989) have been published (Mollhoff, 2001).

There are several magazines to choose from. One of the best bird magazines for amateur birders is *Birder's World*, a beautifully illustrated monthly magazine (circulation 74,000) published by Birder's World Inc 44 E. 8th St Suite 410 Holland, Michigan 49423-3502. The monthly *Wild-Bird Magazine* (circulation 160,000) is generally similar in format to *Birder's World*, and is published at P.O. Box 57900, Los Angeles, CA 90057. Another attractive magazine is the bimonthly *Birding*, published for members by the American Birding Association (A.B.A.), Box 6599, Colorado Springs, CO 80904. This magazine concentrates on articles dealing with the identification of hard-to-identify species and describes favorable locations for birding. As noted earlier the A.B.A. also sells a large variety of bird-oriented books, optical equipment, and recordings of natural sounds, with member discounts. The bimonthly *Bird Watcher's Digest*, published in Marion, Ohio (circulation 82,000), is produced in a format similar to *Reader's Digest*, and has less in the way of colored illustrations but many interesting articles. An attractively illustrated quarterly magazine, *The Living Bird*, is sent to members of the Laboratory of Ornithology, 159 Sapsucker Woods Road, Ithaca, NY 14850. It is similar to *Birder's World* in format, but often includes updates on the laboratory's various research and bird-monitoring programs, such as Project Feederwatch, Project Tanager, and the Nestbox Network. Members of the National Audubon Society can subscribe to the bimonthly *Audubon Society Field Notes* (previously titled *American Birds*), which summarizes seasonal bird sightings across North America, and also publishes results of the annual Christmas Count bird surveys.

For more serious-minded readers there are several organizations and related journals from which to choose. In Nebraska one may wish to join the Nebraska Ornithologists' Union (current treasurer Elizabeth Grenon, 1409 Childs Rd. East, Bellevue, NE 68005; annual dues $15.00), and receive the quarterly *Nebraska Bird Review*, which emphasizes Nebraska bird species and populations. Its homepage website is www.noubirds.org/Default.aspx. There are also several national organizations for serious birders, such as American Ornithologist's Union (P.O. Box 1897, Lawrence, KS 66044-8897, dues $40.00 annually). The AOU publishes *The Auk*, a very technically written quarterly journal containing research papers. Other technical bird journals of national importance are *The Condor, The Wilson Bulletin*, and the *Journal of Field Ornithology*.

Vernacular and Technical Names of Birds

In scanning any field guide or other reference on birds, the reader will soon encounter two sets of names. One will be the English vernacular or "common" name, such as American robin. Such names (often capitalized in bird books and magazines), are the official names given the species by a group such as the American Ornithologists' Union or the American Birding Association. Such uniform names are needed to avoid confusion, such as that caused by having several commonly used names for a single species such as the red-tailed hawk (e.g. the eastern red-tailed hawk, western red-tail, Harlan's hawk, Krider's hawk, Fuertes' red-tail and even, in rural areas, the inappropriate name "chicken hawk"). Occasionally there are necessary changes in vernacular names, as when what had been considered a single species (such as the previously recognized northern oriole) is "split" into two species (namely, the Baltimore and Bullock's orioles). Conversely, what had been two or more previously recognized species (such as the slate-colored, white-winged, gray-headed and Oregon juncos) are "lumped" into a single species (the currently recognized dark-eyed junco). This may cause some confusion to readers, but is sometimes necessary in order to keep the vernacular names of birds in line with current ornithological research.

Even more confusing to most readers are the "scientific" or technical names of birds. These Latin or Latinized names are needed for scientists around the world who speak diverse national languages to be able to communicate effectively. Thus, the European robin is known as the Rotkelchen in Germany and the rouge-gorge in France, but is recognized by scientists as *Erithacus rubecula* in all these countries. Even in English-speaking countries potential confusion exists. Thus, the European robin is a quite different species from the American robin (*Turdus migratorius*), and in Australia the so-called magpies (*Gymnorhina*) are totally different from the magpies (*Pica*) of North America. To avoid such confusion and to provide a basis for a universal nomenclature, scientists have given all of the "kinds" of animals and plants names. These names are a combination of a general or "generic" name (singular = genus, plural = genera) and a specific name (the species, which is the same in singular and plural). The combination of a generic name and specific name represents a unique two-part or "binomial name," such as *Turdus migratorius* for the American robin and *Erithacus rubecula* for the European robin. Of these two components, the generic name comes first and is always capitalized, and the specific name comes second and is never capitalized. Sometimes a third name is added, which designates a recognizable geographic race or subspecies. Like the species, this name is never capitalized and sometimes is exactly the same as the species' name, as in the eastern race of the American robin *Turdus migratorius migratorius* and its western race *Turdus migratorius propinquus*. Such triple names are called trinomials.

There may also be plumage variations that are not given formal recognition in the scientific name, such as genetically variable plumage morphs (commonly but less appropriately also called "phases"). Thus the "blue goose," a genetically controlled plumage morph of the lesser snow goose is not now recognized as a separate subspecies, although it once was. Occasionally distinctive subspecies or morphs are given separate but unofficial vernacular designations, such as the greater and lesser races of the snow goose and sandhill cranes, or the pale-colored "Krider's" variant morph of the red-tailed hawk, sometimes considered a distinct Great Plains subspecies.

At "higher" levels of scientific nomenclature representing categories "above" that of the genus, one encounters progressively larger classification categories containing varied numbers of species and genera, such as the subfamily (which consistently ends in the suffix "inae"), the still larger family (which always ends in "idae"), and the even larger order (which generally ends in "iformes"). Although learning scientific names may seem daunting, at least by knowing that all the species within a single genus are believed by scientists to be close relatives, some understanding of basic bird relationships may be gained. The same applies to genera within a family or families within an order.

National and international committees on biological nomenclature make recommendations as to the most appropriate sequence for listing all such categories and naming their component subdivisions. Complete agreement on these matters and their adoption has yet to be reached by the appropriate committees of all the world's ornithological organizations. Nevertheless, the familiar so-called "Wetmore" (after its originator, Alexander Wetmore) sequence, beginning with flightless birds and loons, and ending with the sparrows and finch-like birds, is generally used in North America and is followed by most field guides and species lists.

Backyard Birding Opportunities

For those persons not interested in traveling far afield to do their bird-watching many opportunities exist for backyard observations. Feeding wild birds is an increasingly popular pastime, and usually allows for close-up viewing of many species, especially during winter. Most feeders allow for seeds such as sunflower seeds (for cardinals and other sparrow-like birds with crushing bills); house sparrows can be deterred from using such feeders by attaching free-hanging monofilament lines around their perimeters. There are also feeders designed to hold small seeds such as thistles (suitable for goldfinches and other small-billed finches) and sugar-water dispensers for hummingbirds. Few hummingbirds stop in Nebraska except along the Missouri River, but once they start to visit a feeder they are likely to remember its location from season to season. Several good books on bird feeding exist, such as Wild About Birds: The DNR Bird Feeding Guide by C. L. Henderson, published in 1995 by the Minnesota Dept. of Natural Resources, St. Paul, MN. Project Feederwatch of Cornell's Laboratory of Ornithology (address given earlier) provides a way of contributing useful information on backyard birds to a national database.

Just as important as regularly maintained feeders is a clean water source, such as a birdbath or pool. Heated birdbaths, fountains and pools that provide a source of constantly dripping or flowing water are especially attractive to birds. Moving water provides a seemingly irresistible attraction for many birds, especially during migration periods. Although "drippers" and "misters" (which spray a fine mist on nearby leaves) are commercially available, simple but effective drippers can be easily made by using a gallon or two-gallon jug, poking one or two tiny holes in its lid, filling it with water, and hanging it above a bird-bath, so the water slowly drips down to the basin below.

Different species (such as hummingbirds) seem to be attracted to misters rather than those favoring drippers, but both are very effective as bird attractants.

Nest boxes offer an additional way of attracting birds during the breeding season. These include not only the traditional wren houses, but also (especially on acreages) bluebird houses, owl houses, wood duck houses, and other nest sites for hole-nesting birds. Although wren houses can be readily purchased, these larger houses often have to be made. Various books on attracting birds to artificial nest sites, such as How to Attract Birds, published by Ortho Books and often sold in garden supply houses, will provide directions. The Wild Bird Habitat Stores of Lincoln, Nebraska and elsewhere, usually have this or similar publications, and also offer a wide array of bird foods and other reference materials. It is important to know that attracting house wrens to one's property will likely cause severe nest losses to other small bird species, especially cavity nesters such as bluebirds. This is a result of the wrens' tendency to take over other birds' nests, often piercing their eggs or even killing the young or adults. Because of this, attracting house wrens on one's property is discouraged, especially if bluebirds, chickadees, tree swallows and similar cavity-nesting species are desired (Radunzel et al. 1997).

Brushpiles of dead branches and twigs are also attractive to many small birds, since these piles offer protection from the cold and from certain predators, such as accipiter hawks. Likewise, dead trees, although they may be an eyesore or potential hazard on a small lot, are favored by cavity-nesters, and on large acreages can be very attractive. An excellent habitat-related booklet is *Landscaping for Wildlife*, published by the Minnesota Dept. of Natural Resources, 500 Lafayette Rd, Box 7, St. Paul, MN 55155 (800-657-3757). This department has also published Woodworking for Wildlife, with many nestbox designs and construction diagrams. Both are collectively available for $9.95.

Monitoring Bird Populations

Bird-banding data have proven abundantly that birds often have rather short lives; a five-year-old American robin is unusual, and a ten-year-old is almost unheard of. A few groups of birds, generally larger species that are effectively protected by habitat or law, such as many seabirds, swans, cranes, and some others, may have lifespans that approach those of humans. Yet, accidents, diseases, predators, starvation and inclement weather all take their toll on wild bird populations, regardless of human efforts to help. In many areas of North America the brown-headed cowbird, a "brood parasite" that lays its eggs in other species' nests and reduces the breeding success of its unwitting hosts, has had devastating effects on many native songbirds as it has expanded its range out of its native Great Plains. Although adapted to prairie and forest edge habitats, forest fragmentation caused by lumbering, road construction, and similar activities has placed many forest-adapted species at risk to such parasitism. Based on Breeding Bird Surveys done between 1966 and 2003, the breeding species that are most rapidly increasing in Nebraska include the wild turkey, Eurasian collared-dove, Canada goose, and Cooper's hawk, whereas the most declining grassland-adapted species are the greater prairie-chicken, grasshopper sparrow, short-eared owl, field sparrow, lark sparrow, loggerhead shrike, long-billed curlew and chestnut-collared longspur. Grassland-adapted species have declined greater nationally than have any other of the ecological categories recognized by the Fish and Wildlife Service, with 75 percent of the grassland species undergoing population declines, owing to losses in native vegetation through conversion to croplands. The four most common breeding species in Nebraska are the western meadowlark, house sparrow, mourning dove and red-winged blackbird; all appear to be declining. State and national population trend estimates of most North American birds can be found at: www.mbr-pwrc.usgs.gov.

Monitoring populations of legally hunted species is the responsibility of national conservation groups such as the U.S. Fish and Wildlife Service, but little effort is made by such groups to monitor populations of songbirds. Here the amateur birder can actively assist, by participating in annual Christmas Bird Counts or Breeding Bird Counts (National Audubon Society), doing Breeding Bird Surveys during June (Biological Resources Division, CSGS), helping with systematic counts at bird feeders (Project Feederwatch of the Laboratory of Ornithology), or maintaining long-term records of the bird populations of a specific location. The Office of Migratory Bird Management of the C.S. Fish and Wildlife Service sponsors a Partners in Flight program that concentrates on needs for monitoring and conserving populations of migratory birds; its activities include organizing workshops, educational programs and bird-related outdoor activities. In conjunction with this program, The Nebraska Game and Parks Commission participates in the International Migratory Bird Day in early May. This often occurs during the same weekend as the Audubon Society's annual Birdathon weekend, which encourages a variety of conservation and bird-awareness activities at the peak of spring migration. Bird-banding activities go on at various nature centers such as at Fontenelle Forest in Bellevue, Neale Woods in Omaha, and Lincoln's Pioneers Park Nature Center. Volunteers are sometimes allowed to become trained in helping to remove captured birds from nets or traps, prior to their banding. Holding a wild bird and releasing it again is a rare and joyful experience for many people, and serves a valuable scientific purpose as well.

Bird diseases, such as the bacterial-carried eye disease that mainly affects the vision of house finches in eastern and central states, and the outbreaks of fowl cholera that regularly occur each spring among waterfowl in south-central Nebraska, are some of the conspicuous avian mortality factors that can be observed and documented fairly easily. For example, persons observing at bird feeders have helped to document the spread of eye disease among house finches in Midwestern and eastern states. Accidents such as deaths caused by birds flying into windows can be avoided or at least ameliorated by affixing parallel strips of conspicuously colored tape along the inner surface of the window (silhouette cutouts of owls or hawks do little good), or by allowing similar plastic strips to hang down freely from the tops of window sills. Free-ranging cats kill millions of wild birds every year; declawing pet cats helps to reduce such needless mortality.

Many states have raptor rehabilitation centers that try to heal and release wild hawks and owls that have met with accidents or been shot by thoughtless gun-owners. In Nebraska the Wachiska chapter of the National Audubon Society in

Lincoln is the state headquarters of these important activities, and volunteers are always welcome (4547 Calvert St. Suite 10, Lincoln 68506, 402/486-4846). The Wachiska chapter also organizes birding outings, Christmas Bird Counts, rare bird alerts, bird-identification classes, prairie preservation and appreciation activities, and related conservation programs. In Omaha similar activities are performed by the Audubon Society of Omaha, (Center Mall, 1941 S. 42nd St #501, Omaha 68105, 402/342-1345), and in Kearney the Big Bend chapter (P.O. Box 1575, Kearney 68848) helps organize the annual spring river conference each March. There is also a Wildcat Audubon chapter in Scottsbluff, a Loess Hills chapter in Sioux City (P.O. Box 5133, Sioux City 51102), and inactive chapters in Grand Island and the Valentine/Ainsworth area. Phone numbers or contact people for these smaller groups can be obtained from the state office: Audubon Nebraska, P.O. Box 117, Denton, NE (402/793-2301).

The Audubon Society also sponsors the annual Rivers & Wildlife Conference during mid-March in Kearney. Other popular spring bird-related activities occurring during the same month are the Wing Ding celebration at Clay Center in early March, celebrating the waterfowl migration in the Rainwater Basin, and the Wings over the Platte activities in Grand Island during the latter part of March. Establishing and patrolling bluebird nestbox trails has had a major effect on restoring the eastern bluebird as a breeding species in Nebraska, and has also similarly benefited tree swallows. The organization Bluebirds Across Nebraska is the sponsor of this program throughout the state, and in Lincoln it is coordinated through the local chapter of the Audubon Society (see address above).

Nationally, nature-related tourism has been growing at a rate of 30% annually, with 76.5 million Americans now viewing wildlife, and 24.7 million observing and/or feeding birds. By comparison, there are 35.6 million American anglers and 14.1 million hunters. Nature tourism and recreation now generates over $20 billion in economic activity and results in 234,000 jobs in 1991, Americans spent $4.4 billion for food and lodging to enjoy wildlife for non-consumptive purposes.

Bird-finding Areas and Birding Information Sources in Nebraska

This bird-finding guide was first assembled in 1996 to try to fill the gap in available information on potential locations for bird-watchers in Nebraska. Since then, a guide to wildlife watching in the state has been published (Krue, 1997), and a special issue of NEBRASKAland on state bird-watching ("Birding Nebraska" by Jon Farrar) was published in 2004. The following section of this guide represents an effort to identify and describe nearly all of the 400+ public-access sites in the state of Nebraska of special interest to birders and other naturalists. It includes all of the state's national wildlife refuges, federal waterfowl production areas and national monuments. It also includes all the state parks, most state historical parks and state recreation areas, and all of the larger wildlife management areas. Municipal and county parks are usually not included unless they are of particular biological interest. Maps showing specific locations (and approximate geographic limits in most cases) for over 300 major birding sites in 43 counties of the state are also provided in this guide. The remaining 50 unmapped counties have few or no public access sites of special interest to naturalists, and those sites mentioned are accompanied with location descriptions that should suffice for finding them with the use of a state highway map. Maps of CRP lands that are open to public use can be accessed through the hunting section of http://outdoornebraska.ne.gov.

As noted earlier, a web-based and searchable version of this guide is now available at: www.nebraskabirdingtrails.com. The website is sponsored in part by the Nebraska Bird Partnership: www.nebraskabirds.org. This group is a non-profit consortium of many conservation and bird-oriented groups and agencies that have the goal of providing a coordinated, science-based landscape approach to voluntary land stewardship for conserving, improving and expanding habitats for all birds. The Partnership recognizes the importance of sustainable natural communities and a common need for effective communication, education and a broad appreciation for the diversity of bird habitats and bird species in Nebraska, and sponsors periodic symposia.

Sets of individual county road maps (scale 1" or ¼" per mile) are available from the Nebraska

Roads Department (1500 Highway N-2, Lincoln 68509; 402/471-4567). An atlas of 79 topographic maps of the entire state (scale 1:200,000, or ca. 1/3" per mile) is available for $16.95 in the *Nebraska Atlas and Gazetteer*, published by DeLorme, PO Box 298, Freeport, ME 04032 (207/865-4171). This atlas also shows state parks and recreation areas, national lands, campgrounds, wildlife viewing areas, fishing, hiking and other attractions. A similar atlas having maps (scale ¼" per mile) of all of Nebraska's 93 counties, with descriptions of hunting, fishing, camping and related outdoor attractions is the Nebraska Sportsman's Atlas, available for $18.75 from Sportsman's Atlas Company, PO Box 132, Lytton, Iowa 5056 (800/568-8334). The DeLorme atlas is especially good for information on elevations, river drainages and forested areas, as well as for latitudinal/longitudinal estimate; the Sportsman's maps (based on state Roads Department maps) are better for obtaining road and recreational site information. Nearly all the sites described here are now mapped on the Nebraska Birding Trails website: www.nebraskabirdingtrails.com.

Nebraska's rivers are publicly owned, but the adjoining shorelines and islands may be in private ownership. Birding from a canoe is possible on several rivers (Niobrara, Dismal, Calamus, Missouri, Platte, Republican), but access points are often limited. The longest stretch of river ideal for canoe-based birding is the 76-mile section of the Niobrara designated as a National Scenic River. The Dismal and Calamus also offer good canoeing but have limited access points. Some refuge lakes at Valentine National Wildlife Refuge also offer wonderful birding opportunities from a canoe. Those at Crescent Lake N.W.R. are closed to such activities.

The Nebraska Division of Travel and Tourism (301 Centennial Mall S., Lincoln, NE 68509) can provide free information on general tourist attractions (402/471-3441 or 800/742-7595; from out of state call 800/228-4307). Their website is www.visitnebraska.gov, and their contact e-mail addresses are listed at www.visitnebraska.gov/tourism-contacts. Tourism information and free state highway maps are also available at most Interstate rest areas. Road information can be obtained in-state by calling 800/906-9069; if out-of-state call 402/471-4533. Their materials list all tourist accommodations in the state; where tourist accommodations are mentioned in this summary as available, they refer to hotels or motels, not campgrounds. The emergency highway help line is 800/525-5555. State recreation areas usually offer more highly developed recreational facilities and modern camping. Annual park entry permits or more information can be obtained from the Parks Division, Game and Parks Commission, PO Box 30370, Lincoln, NE 68503, (800/826-PARK). District offices are also located in Alliance (308/762-5605), Bassett (402/684-2921), Norfolk (402/370-3374) and North Platte (308-535-8025). The Nebraska Game and Parks Commission's "Land Atlas" is a web-based map of areas owned or leased by the Commission in each Nebraska county and that are available for public access and recreational use. Their website also has some very useful information for birders and links to other websites: http://outdoornebraska. ne.gov. The Patuxent Wildlife Research Center has also a valuable website: www.pwrc.usgs.gov. The Patuxent Bird Identification InfoCenter has photos, songs, videos, maps and life history information on most North American birds: www.mbr-pwrc. usgs.gov/id/framlst/infocenter.html.

Rare birds seen in Nebraska should be described and reported to the Nebraska Ornithologists' Union (NOU) for documentation in their quarterly journal Nebraska Bird Review. The NOU also publishes periodic newsletters for members. It has spring and fall meetings, usually in September and May, at locations favorable for seasonal birding. The NOU website (address provided earlier, in "Reference Materials") has a great deal of useful bird information. NEBIRDS is a listserv for reporting recent Nebraska bird sightings and sharing bird information. It is accessible by free subscription by going to the following website: http://groups.yahoo.com/group/NEBirds.

Click on the link labeled "Join This Group," and provide the requested information. You will have to become a Yahoo member before you can join any group. When you fill out the Yahoo membership information uncheck the boxes relating to receiving information from Yahoo if you want to avoid spam from Yahoo advertisers. Up-to-date information on the sightings of unusual birds in Nebraska (and other states or regions) is also available on the Internet via the general birding website http://birdingonthe.net.

The following guide section is organized into four general regions (Far Western, West-central, East-central, and Eastern, and within each of these four regions individual counties are sequentially discussed in a generally west-to-east and north-to-south sequence, as indicated on the accompanying map of Nebraska counties (see middle map on p. 17). Most federally owned birding areas in Nebraska consist of national historic sites, national wildlife refuges and national monuments. Federally owned areas also include waterfowl production areas (WPAs). State-owned sites include state parks, state recreation areas (SRAs), and wildlife management areas (WMAs). No permit is needed to enter WPAs or WMAs, but annual (or daily) state park entry permits are needed for SRAs and all state parks and state historical parks. One regional national wildlife refuge (DeSoto) also charges a daily entry fee ($3.00). All state wildlife management areas offer free, unrestricted birding or other nature-related opportunities. They usually provide only primitive camping facilities, or none, and most are open to seasonal hunting and fishing.

Ecologically based regions of Nebraska, as shown at the county level (bottom map, p.17), provide a generalized view of the state's major eco-geographic regions. Habitat descriptions have not yet been entirely standardized for Nebraska but the categories shown on page 102 provide for convenient, uniform descriptions. These categories are largely based on a system proposed for Nebraska by G. Steinauer and S. Rolfsmeier (2003).

The Nebraska Tourism Department has traditionally divided the state into several tourism-based regional units including the western "Panhandle Region," a north-central "Sandhills Region," a northeastern "Lewis & Clark Region," a southwestern "Prairie Lakes Region," a south-central "Frontier Trails Region," a southeastern "Pioneer Country Region," and a "Metro Region" encompassing greater Omaha and Lincoln. The Nebraska Game & Parks Commission also uses these regions. The Tourism Department's website (www.visitnebraska.gov) is useful for finding sleeping, eating facilities and recreational activities around the state and interstate rest areas as well. In addition, the tourist attractions provide free hardcopy state tourism guides.

Nebraska has numbered its 93 counties relative to historic population size (1 = largest, 93 = smallest). The preliminary numbers on car licenses correspond to these numbers, except for a few eastern counties (Lancaster, Douglas & Sarpy), which have replaced these numbers with random three-letter combinations. The top map on page 7 shows the county names. The state's emergency highway help line is 1-800-525-5555. Its cellular phone hotline is *55 (star 55). Road reports are available in-state via 1-800-906-9069. From Omaha dial 553-5000, and from out of state dial (402) 471-4533. The State Patrol Citizen Band Call Letters are KNE 0911. The statewide Crime Stoppers number is 1-800-422-1494.

Maps on following pages:

Page 17. Maps of Nebraska counties (top), general birding regions used in this book (Roman numerals) and numbering system for counties located in each region (middle), and county-based ecological regions of Nebraska (bottom).

Page 18. Map of Nebraska State Parks and State Recreation Areas (courtesy Nebraska Game & Parks Commission), plus locations of various federal lands and private (The Nature Conservancy and National Audubon Society) preserves.

Top map — Nebraska counties:

SIOUX, DAWES, BOX BUTTE, SHERIDAN, CHERRY, KEYA PAHA, BOYD, KNOX, CEDAR, DIXON, DA-KOTA, BROWN, ROCK, HOLT, ANTELOPE, PIERCE, WAYNE, THURSTON, SCOTTS BLUFF, MORRILL, GARDEN, GRANT, HOOKER, THOMAS, BLAINE, LOUP, GARFIELD, WHEELER, MADISON, STANTON, CUMING, BURT, BANNER, ARTHUR, McPHERSON, LOGAN, VALLEY, GREELEY, BOONE, PLATTE, COLFAX, DODGE, WASHINGTON, KIMBALL, CHEYENNE, DEUEL, KEITH, NANCE, SHERMAN, HOWARD, MERRICK, POLK, BUTLER, SAUNDERS, DOUGLAS, SARPY, LINCOLN, PERKINS, DAWSON, BUFFALO, HALL, HAMILTON, YORK, SEWARD, LAN-CASTER, CASS, OTOE, CHASE, HAYES, FRONTIER, GOSPER, PHELPS, KEARNEY, ADAMS, CLAY, FILLMORE, SALINE, JOHN-SON, NE-MAHA, DUNDY, HITCHCOCK, RED WILLOW, FURNAS, HARLAN, FRANKLIN, WEBSTER, NUCKOLLS, THAYER, JEFFER-SON, GAGE, PAWNEE, RICHARD-SON

Middle map — Regional (Roman numerals) and county numbering system:

I. II. III. IV.

Bottom map — County-based ecological regions of Nebraska:

Pine Ridge, Central Niobrara Valley, Missouri Valley, Sandhills, Tallgrass Prairie, Shortgrass Prairie, Central Platte Valley, Sandsage Prairie, Rainwater Basin/Mixed-grass Prairie, Republican Valley

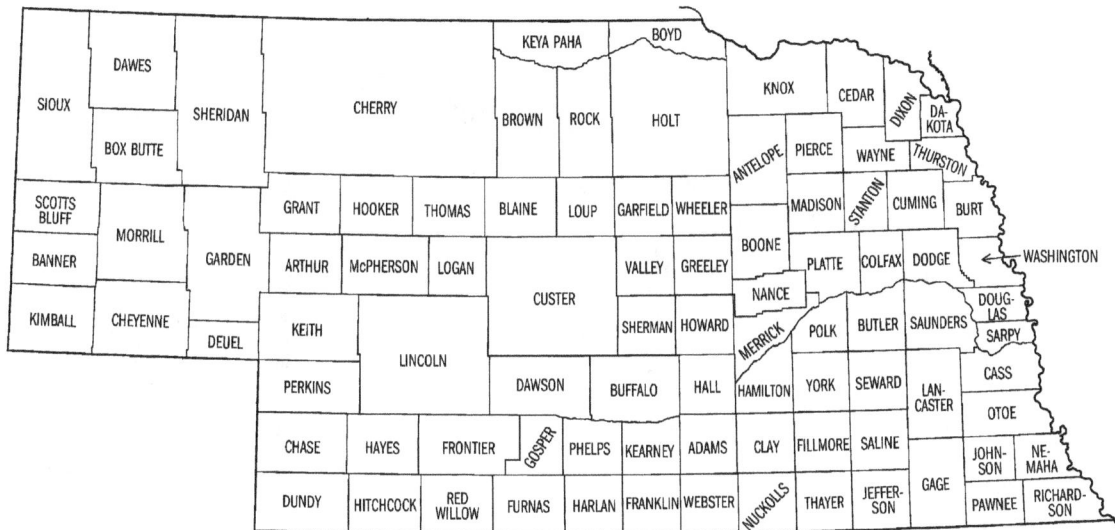

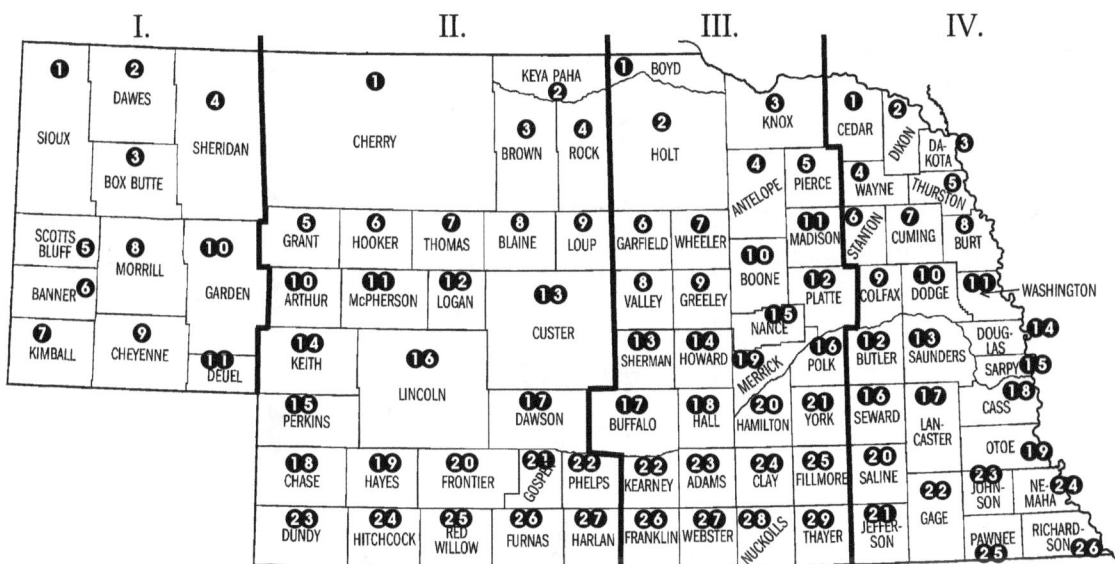

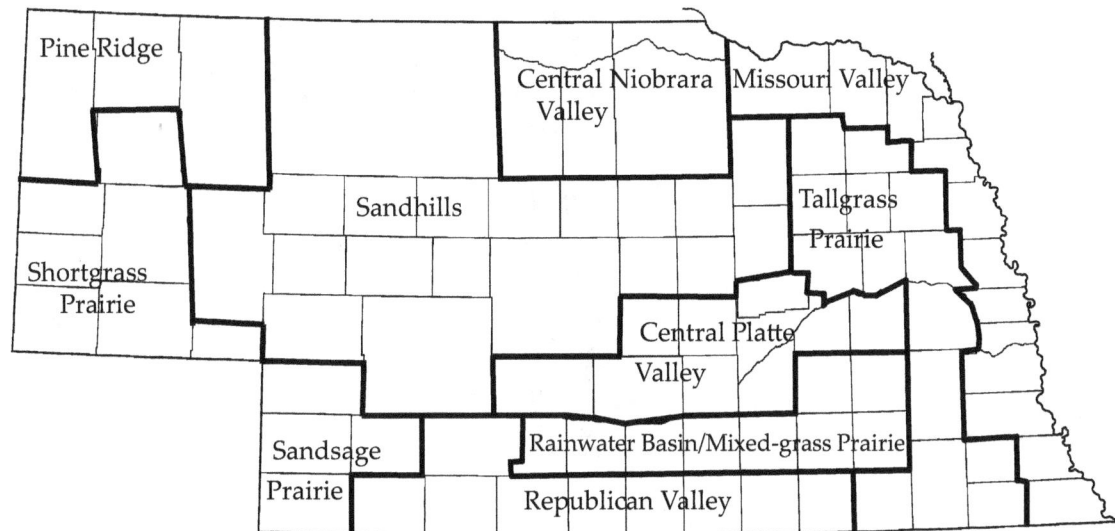

Top: Nebraska counties. *Middle*: Regional (Roman numerals) and county numbering system used in this book. *Bottom*: County-based ecological regions of Nebraska.

Nebraska State Parks and State Recreation Areas

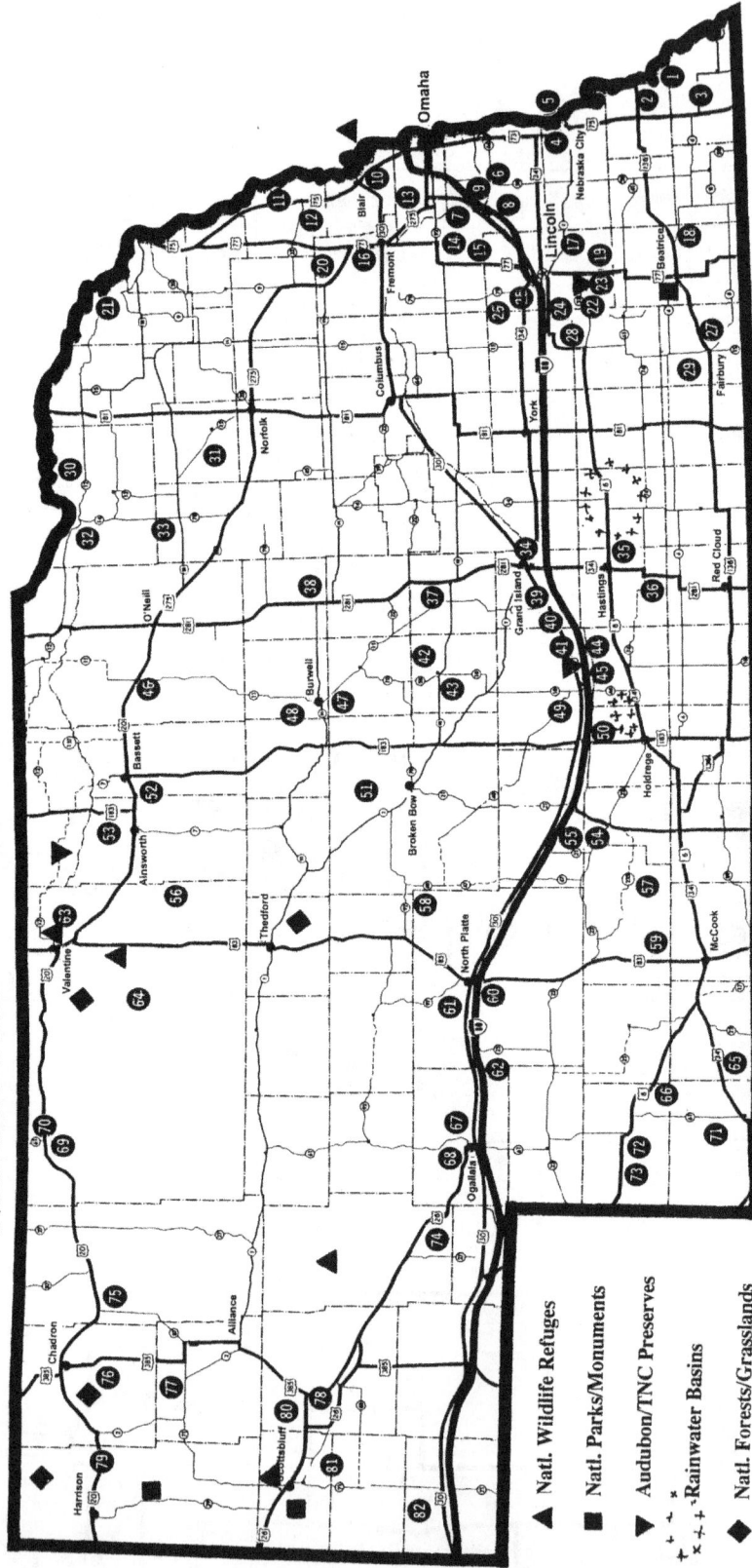

▲ Natl. Wildlife Refuges

■ Natl. Parks/Monuments

▼ Audubon/TNC Preserves

✕ ✕ ✕ Rainwater Basins

◆ Natl. Forests/Grasslands

1. Indian Cave State Park
2. Brownville SRA
3. Verdon Lake SRA
4. Arbor Lodge State SHP
5. Riverview Marina SRA
6. Louisville Lakes SRA
7. Eugene T. Mahoney SP
8. Platte River State Park
9. Schramm Park SRA
10. Fort Atkinson SHP
11. Pelican Point SRA
12. Summit Reservoir SRA

13. Two Rivers SRA
14. Memphis Lake SRA
15. Pioneer SRA
16. Fremont Lakes SRA
17. Wagon Train SRA
18. Rockford Lake SRA
19. Stagecoach SRA
20. Dead Timber SRA
21. Ponca State Park
22. Bluestem SRA
23. Olive Creek SRA
24. Conestoga SRA

25. Pawnee SRA
26. Branched Oak SRA
27. Rock Creek Station SHP & SRA
28. Blue River SRA
29. Alexandria Lakes SRA
30. Lewis & Clark SRA
31. Willow Creek SRA
32. Niobrara State Park
33. Ashfall Fossil Beds SHP
34. Mormon Island SRA
35. DLD SRA
36. Crystal Lake SRA

37. North Loup SRA
38. Pibel Lake SRA
39. Cheyenne SRA
40. War Axe SRA
41. Windmill SRA
42. Sherman Reservoir SRA
43. Bowman Lake SRA
44. Fort Kearny SRA
45. Fort Kearny SHP
46. Atkinson Lake SRA
47. Fort Hartsuff SHP
48. Calamus SRA

49. Union Pacific SRA
50. Sandy Channel SRA
51. Victoria Springs SRA
52. Long Pine SRA
53. Keller Park SRA
54. Johnson Lake SRA
55. Gallagher Canyon SRA
56. Long Lake SRA
57. Medicine Creek SRA
58. Arnold Lake SRA
59. Red Willow Reservoir SRA
60. Lake Maloney SRA

61. Buffalo Bill Ranch SHP & SRA
62. Sutherland Reservoir SRA
63. Smith Falls State Park
64. Merritt Reservoir SRA
65. Swanson Reservoir SRA
66. Enders Reservoir SRA
67. Lake Ogallala SRA
68. Lake McConaughy SRA
69. Cottonwood Lake SRA
70. Arthur Bowring Sandhills Ranch SHP
71. Rock Creek SRA

72. Champion Mill SHP
73. Champion Lake SRA
74. Ash Hollow SHP
75. Walgren Lake SRA
76. Chadron State Park
77. Box Butte SRA
78. Bridgeport SRA
79. Fort Robinson State Park
80. Lake Minatare SRA
81. Wildcat Hills SRA
82. Oliver Reservoir SRA

The Far Western Region: Pine Ridge Country

This beautiful part of Nebraska, its geographic "Panhandle," is largely a ridge-and-canyon region, interspersed with High Plains topography and steppe vegetation. It is the land that Crazy Horse died to try to protect for his people, the Oglala Sioux or Lakotas, and one laced with the bitter history of these people and Cheyennes as they vainly fought to maintain their sacred lands. The pine-covered hills and escarpments remind one of the Black Hills, and about 3.5 percent of the region's land is covered by wooded habitats (Nebraska Game & Park Commission, 1972). Several pine-adapted species that are common in the Black Hills breed only in the northwestern corner of Nebraska, such as Lewis' woodpecker, pinyon jay, dark-eyed junco, western tanager, yellow-rumped warbler, Swainson's thrush, solitary vireo, red-breasted nuthatch and red crossbill. Some of these same species as well as the violet-green swallow, white-throated swift and pygmy nuthatch also occur in the pine forests of the Scottsbluff area and the Wildcat Hills. However, the canyon-adapted cordilleran (previously called "western") flycatcher is mostly limited to Sowbelly Canyon, Sioux County but has recently been reported at a few other sites.

The Panhandle region also has over 5.6 million acres of grassland, which supports a few quite localized short-grass or arid plains species such as the McCown's and chestnut-collared longspurs and mountain plovers. A few sage-adapted species, including the sage thrasher and the Brewer's sparrow, are also possibilities. It is also a land rich in the fossil remains of early Cenozoic mammals, as well as an eight-million-year-old fossil bird bone that appears to be identical to that of a modern sandhill crane. This would make the sandhill crane the most archaic of all known extant birds, and provides another reason for considering it a very special bird species. Sandhill cranes by the tens of thousands still pass through this region each spring and fall, but their major migratory pathway lies to the east, in the central Platte Valley. Typical open-country Panhandle birds include the prairie falcon, ferruginous hawk, golden eagle, merlin, Say's phoebe, western wood-pewee, Cassin's kingbird, pinyon jay, rock wren, and McCown's longspur, while the mountain bluebird, yellow-rumped warbler, western tanager, and pygmy nuthatch are more woods adapted.

The major birding attractions in the Panhandle include the biologically diverse and scenic Pine Ridge area. A bird checklist encompassing "northwest Nebraska," is based on Richard Rosche's 1982 report on 324 species reported in four northwestern counties (see supplement). Secondly, the area around Lake McConaughy has one of the very few local bird lists exceeding 325 reported species for any site north of Mexico. About 75 miles northwest is Crescent Lake National Wildlife Refuge, a wilderness refuge in the western Sandhills, having the second-largest local bird list for the state, with 273 species (also summarized in the supplement). To the north of Crescent Lake, in northern Garden County and southern Sheridan County, are hundreds of highly saline Sandhills marshes that often abound with waterfowl, shorebirds, and marshland birds.

1. Sioux County (Map 1; Recently Burned Areas Shaded)

Sioux County is in the heart of the Pine Ridge region, an area of ridge-and-canyon topography that is a southern outlier of the Black Hills region of South Dakota, and is a north-facing escarpment largely covered by ponderosa pine forest and streamside deciduous forests, totaling some 68,000 acres. As such, it has several species that occur rarely if at all elsewhere in Nebraska, such as the cordilleran flycatcher and plumbeous vireo. There are also more than a million acres of short-grass plains, much of which are included in the Oglala National Grassland, and which support a typical high plains avifauna (Boyle & Bauer, 1994). There are tourist accommodations at Harrison.

A. Federal Areas

1. Oglala National Grassland (Map locations 1, 3). Area 93,344 acres. The area around Toadstool Park offers Brewer's sparrows, sage thrashers, long-billed curlews, Swainson's and ferruginous hawks, and chestnut-collared longspurs (Map location 3). Horned larks, western meadowlarks and lark buntings are common breeders in this vast region, which extends into Dawes County. Large prairie dog towns occur north of Montrose. Ferruginous hawks and golden eagles are regularly present.

For information contact the Forest Service office at HC 75, Box 13A9, Chadron NE 69337 (308/432-4475). www.fs.usda.gov/nebraska

2. Soldier Creek Wilderness (Map location 7). Area 9,600 acres. This is a large roadless area that has an extensive hiking trail network, as well as bridle trails. Water must be carried in, and facilities are lacking. Much of the area was burned in a 1989 fire. An eight-mile loop trail over ridges and canyons has its trailhead at the picnic area. For information contact the Forest Service office mentioned above. www.fs.usda.gov/nebraska

3. Agate National Monument. Not shown; see a state highway map for location. Includes nearly 2,000 acres of shortgrass plains. No official bird checklist is yet available, but 156 species have been reported for the site, including ferruginous hawk, mountain plover, burrowing owl, saw-whet owl, white-throated swift, Cassin's kingbird, pinyon jay, Townsend's warbler, western tanager, black-headed grosbeak, lazuli bunting, and three species of longspurs including both McCown's and chestnut-collared. For information call the National Park office at 308/668-2211. www.nps.gov/agfo

4. Toadstool Geologic Park (Map location 2). This area of badlands (ca. 300 acres) within the Oglala National Grassland supports rock wrens, Say's phoebes, golden eagles and prairie falcons, and sometimes also gray-crowned rosy finches during winter. A one-mile loop trail through part of the

park that begins at the picnic area should turn up rock wrens and other topography-dependent birds. Water is at a premium here, and a canteen should be carried in hot weather. A small campground is present (call 308/432-4475). www.fs.fed.us/wildflowers/regions/rockymountain/ToadstoolGeoPark/index.shtml

5. Nebraska National Forest, Pine Ridge District (Map location 10). This area comprises about 51,000 acres, with most holdings in Dawes County (see Dawes County).

B. State Areas

1. Fort Robinson State Park (Map location 9). Area 22,000 acres. Although still providing good pine habitat, a forest fire in 1989 destroyed much of the best sections of the park, which does offer lodging and eating facilities (Rosche, 1990). A nesting area for white-throated swifts occurs 6 miles west of headquarters (Pettingill, 1981). For information call 308/665-2900. www.stateparks.com/fort_robinson.html

2. Gilbert-Baker WMA (Map location 4), Area 2,457 acres. This is an area of ridges covered with ponderosa pines, with scattered areas of grassland at the forest fringes. Monroe Creek traverses the area and is a trout stream. Located 3 miles north of Harrison, via an oil-surfaced road. A gravel road going south along the Wyoming border (turn 8 miles west of Harrison) crosses the Niobrara River and passes into ridge-and-valley topography that supports McCown's longspurs, Say's phoebes and rock wrens, as well as Brewer's sparrows, ferruginous hawks and long-billed curlews, plus chestnut-collared longspurs farther south. At about 8 miles south of the turning a road goes east and back to state highway 29 (Rosche, 1990). Hiking trails penetrate the area; for information call 308/668-2211.

3. Peterson WMA (Map location 6). Area 2,460 acres. This area consists of habitats alternating between mature ponderosa pine forests and grasslands in typical ridge-and canyon topography. Two streams bisect the area. There are no camping facilities.

4. James Ranch SRA (Map location 8).

C. Other Areas

1. Sowbelly Canyon (Map location 5). Although privately owned, this area is reached via a county road northeast from Harrison (drive 1 mile north, then turn east and proceed northeast for several miles along Sowbelly Creek). This road enters a narrow canyon and passes through passes through a creekbottom area (Coffee Park) where on-foot birding can be done, about 5 miles from town. Many distinctly western species previously bred here (Rosche, 1990), but forest fires in 2006 engulfed the entire canyon, and many other parts of the Pine Ridge (see outlined areas on Maps 1 & 2), covering 65,000 acres and destroying a large proportion of the region's ponderosa pine forests. Inquire locally before visiting this locality.

2. Monroe Canyon is directly north of Harrison, the lower portion in the Gilbert-Baker WMA (see above), and mostly deciduous forest. The upper part is ponderosa pine forest. Drive down the canyon road, stopping every few hundred yards. The small side canyons are used by cordilleran flycatchers. There is a campground at the bottom of the canyon along Monroe Creek. All the Pine Ridge species are there. If you drive north from the campground entrance about 200 yards a gravel track leads east to an impoundment that has violet-green swallows (Ross Silcock, NOU website).

3. Smiley Canyon. This canyon is reached from Fort Robinson State Park. About a mile west of the Fort take a paved road north that leads through grasslands (look for bison) to an area of ponderosa that was burned a few years ago. Look for Lewis's, black-backed and possibly even three-toed woodpeckers (Ross Silcock, NOU website).

2. Dawes County (Map 2; Recently Burned Areas Shaded)

Dawes County is one of Nebraska's most scenic regions, with nearly 100,000 acres of wooded habitats, and almost 600,000 acres of grasslands within its boundaries. There are tourist accommodations at Chadron and Crawford.

A. Federal Areas

1. Pine Ridge National Recreation Area and Nebraska National Forest, Pine Ridge Unit (Map location 2). Recreation Area 50,803 acres, mostly of ponderosa pine forest and intervening grasslands or farmlands. The topography of this area is often rugged, and the roads may not be in good condition, so it is well to check with the ranger office on US highway 385 before venturing far from the main road. Cattle grazing is permitted here, so attention to gates is needed. There is a 4-mile fairly difficult trail starting at the Iron Horse Road meadow, and a less difficult 3-mile hiking trail with its trailhead at East Ash Road. A fairly difficult 8-mile trail leading to Chadron State Park begins at a gravel road off US highway 85. For more information contact the Forest Supervisor at 270 Pine St., Chadron (308/432-4475) www.fs.usda.gov/nebraska.

2. Oglala National Grasslands (Map location 1. Note that fine lines enclose actual holdings; broad lines show maximum limits of grassland district.) Area 94,394 acres. See Sioux County account. www.fs.usda.gov/nebraska

B. State Areas

1. Fort Robinson State Park (Map locations 5, 8). Area 20,000 acres. See also Sioux County account. State park entry permit required; call 308/665-2900 for information. www.stateparks.com/fort_robinson.html

2. Chadron State Park (Map location 7). Area 801 acres. This is the best place in the region to see Lewis' woodpecker, and it also supports pygmy nuthatches, western tanagers and common poorwills (Rosche, 1990). On the way to the Black Hills lookout watch for Lewis' woodpeckers perched on the tops of snags. At the lookout one should see pinyon jays, yellow-rumped warblers, western tanagers and mountain bluebirds, as well as raptors (Boyle & Bauer, 1994). Pettingill (1981) has described birding opportunities in this park, and provides a list of nesting species. The Spotted Tail hiking trail extends for 8 miles from the park boundary through the Nebraska National Forest, and the Black Hills Overlook trail extends for 4 miles from the park campground. State park entry permit required; call 308/432-6167 for information. http://outdoornebraska.ne.gov

3. Ponderosa WMA (Map location 6). Area 3,659 acres. Located southeast of Crawford, this area is largely covered by ponderosa pine forests, with grasslands on level areas and also some decidu-

ous trees lining Squaw Creek. There is a hiking trail starting at parking area 5 that provides an excellent panorama, and may offer views of such raptors as prairie falcons. National Forest land adjoins the area to the south and southwest. About 10 miles south of Crawford along state highway 2 are ridgetop pine wooded habitats where Cassin's kingbirds are rather easily seen, especially during September (Rosche, 1990).

4. Box Butte SRA (Map location 3). Land area 612 acres. Includes a 1600-acre reservoir. This is an outstanding birding area in the Panhandle; Richard Rosche (personal communication) has observed over 200 species in a 20-year span. Rock wrens, Say's phoebes and ferruginous hawks are among the more interesting western species, and probable eastern breeders include eastern bluebird, eastern wood-pewee, and wood thrush. Small passerines such as warblers and vireos are abundant during migration. http://outdoornebraska.ne.gov

5. Whitney Lake WMA (Map location 4). Area 900 acres. Located 2 miles northwest of Whitney.

3. Box Butte County

Box Butte County is only slightly forested (about 5,000 wooded acres), but has over 300,000 acres of remaining grasslands or farmlands. There are tourist accommodations at Alliance.

A. Federal Areas: None

B. State Areas: None

C. Other Areas
1. Kilpatrick Lake. Located about 20 miles west of Alliance (west on 10th St. for 11 miles, then south 1 mile, west for 5 more miles.) A trail at a sign indicating the Snake Creek Ranch goes left and leads to the dam. This privately owned reservoir (recently low) is a major stopover point for snow geese and a few Ross' geese in spring. The meadows to the south of the dam around Snake Creek support willets, long-billed curlews, Wilson's snipes, eastern meadowlarks and Savannah sparrows, and the drier areas should be scanned for Cassin's sparrows (rare in Nebraska). These birds inhabit sandsage grassland, and their vocalizations help locate them.

4. Sheridan County (Map 3)

Sheridan County has about 50,000 acres of wooded habitats, and nearly 1.2 million acres of grasslands present. It also has over 20,000 acres of surface wetlands, much of which consists of Sandhills marshes. There are tourist accommodations at Gordon and Rushville.

A. Federal Areas. None.

B. State Areas
1. Metcalf WMA. Area 3068 acres. Located about 10 miles north of Hay Springs. It has typical Pine Ridge habitat, which is mostly pine-covered, but with some open grasslands present. There are no camping facilities.

2. Smith Lake WMA. Area 640 acres. Located 20 miles south of Rushville. The area has a 222-acre lake, surrounding marsh and grasslands, and some wooded habitats. There are primitive camping facilities and toilets. Fishing is permitted. Between Lakeside and Rushville, 50 miles apart, excellent birding opportunities exist, and nesting records for the long-eared owl, black-necked stilt, piping plover and even the northern parula have been obtained (Rosche, 1990).

3. Walgren Lake WMA. Area 130 acres. Located near Hay Springs; see state highway map for exact location. There are primitive camping facilities. A great variety of migrant species are attracted to this lake, including such rarities as Sabine's and black-headed gulls, and Townsend's warbler. Just a mile south of Walgren Lake is a prairie dog town with nesting burrowing owls and occasional chestnut-collared longspurs. The latter are more common along the first road going east to the north of the colony (Rosche, 1990).

C. Other Areas
1. Sandhills marshes near Lakeside. (Map location 2). This area, extending west and east from Lakeside on US Highway 2, and north on state highway 250, provides views of many highly alkaline marshes that attract waterfowl such as trumpeter swans and many shorebirds such as breeding black-necked stilts, American avocets, willets, Wilson's phalaropes, and others. Taking the gravel road south from Lakeside takes one

(28 miles, no gas or facilities) through Sandhills country to Crescent Lake National Wildlife Refuge (see Map 3, and Garden County account), and past many wet meadows and very saline marshes (those wetlands north of the dashed line on Map 3) that are highly attractive to shorebirds and waterfowl. Sandhills roads are narrow, hilly and sometimes slippery; careful driving is mandatory. The road (state route 250) north from Lakeside is just as attractive; after about 20 miles an unimproved road going east connects with state route 27 and returns one to US highway 2 at Ellsworth. However, it might be better to backtrack from Smith Lake to Lakeside and make a similar two-way run north from Ellsworth for about 15 miles, where the marshy wetlands peter out.

5. Scotts Bluff County (Map 4)

Scotts Bluff County has over 25,000 acres of wooded habitats, and more than 180,000 acres of grasslands within its boundaries. There are tourist accommodations at Gering and Scottsbluff.

A. Federal Areas

1. North Platte National Wildlife Refuge, including Lake Minatare State Recreation Area (Map location 2). Area 5047 acres. The best part of this refuge is the 500-acre Winters Creek Lake Unit northwest of Lake Minatare, where a marshy lake attracts a large number of migratory and breeding water birds, including western grebes. There is a bird list for the entire refuge (a major basis for the list of North Platte Valley birds in the supplemental checklist). The checklist includes 181 species, with 32 known nesters and 20 additional possible breeders. For information contact the local Fish and Wildlife Service office at 308/635-7851. www.fws.gov/crescentlake

2. Scotts Bluff National Monument (Map location 1). Area 3,000 acres. This famous bluff along the Oregon Trail is capped by ponderosa pine wooded habitats, and has steep sides that are used as nesting sites by white-throated swifts. At least 100 species have been reported for the area, including prairie falcons burrowing owl, common poorwill, pinyon jay, both cuckoos, rock wren, yellow-rumped warbler, Baltimore and Bullock's orioles, blue and black-headed grosbeaks, green-tailed and spotted towhees, three races of dark-eyed juncos, and lazuli bunting (unpublished staff records). There is a 3-mile nature trail leading from the summit parking lot to the visitor center. Three prairie dog towns totaling about 63 acres are present. For information contact the superintendent at Box 27, Gering NE 69341 (308/436-4340). www.nps.gov/scbl

B. State Areas

1. Nine Mile Creek Special Use Area (Map location 3). Area 178 acres. Located north and east of Minatare, and consisting of grasslands, plus a trout stream.

2. Wildcat Hills State Recreation Area and Buffalo Creek WMA (Map location 4). Area 3935 acres. Buffalo Creek WMA consists of typical Wildcat Hills ridge-and-canyon habitats, covered by pines and junipers. It is nearly all wooded, but has a seven-acre pond. Primitive camping facilities are present, and there is a long trail through Buffalo Creek WMA. Pygmy nuthatches nest here, and violet-green swallows are fairly common. Several raptors, such as golden eagles, prairie falcons and several buteos are good possibilities. Pettingill (1981) listed 11 species that should potentially be seen here, including common poorwill and white-throated swift. A nature center is present at the SRA, and a 2-mile nature trail. Red crossbills and red-breasted nuthatches are regular at feeders here. For information call 308/436-3777. www.nebraskabirdingtrails.com

3. Kiowa WMA. (Not shown, located 2.5 miles south of Morrill). Area 540 acres, of which 326 acres are closed during the goose-hunting season as a refuge. There is a large pool at the east end, and seasonal wetlands toward the west. A small prairie dog colony is present, and burrowing owls are regular. About 20 waterfowl species have been reported, including large wintering populations. Nesting birds include American avocet and black-necked stilt. A Nebraska Important Bird Area.

6. Banner County

Banner County is a high plains county that is slightly wooded (under 25,000 acres of wooded habitats), and has about 235,000 acres of grassland habitats. Very little surface water is present. There are no tourist accommodations in the county.

A. Federal Areas: None

B. State Areas
   1. Buffalo Creek WMA. See Scotts Bluff County.

   2. Wildcat Hills SRA. See Scotts Bluff County.
www.nebraskabirdingtrails.com

7. Kimball County (Map 5)

Kimball County is a high plains county with only about 500 acres of wooded habitats, a similar acreage of surface water, and about 180,000 acres of grasslands or farmlands. There are tourist accommodations at Kimball.

A. Federal Areas: None

B. State Areas
   1. Oliver Reservoir SRA (Map location 1). Area 1,187 acres. This reservoir of Lodgepole Creek is an excellent birding location, attracting many migrants during spring and fall, especially warblers. Wilson's snipe have been reported to nest here (west-end marshes), and a population of song sparrows is the only one known for western Nebraska (Rosche, 1994). This is the largest reservoir in southwestern Nebraska, and attracts many migrants in spring (late March to early June) and fall (late August to early November). Western rarities include dusky flycatcher (May & August-September), Cassin's vireo (August-September), and Townsend's warbler (August-October. More than 210 species have been seen here (Stephen Dinsmore, NOU website).
http://outdoornebraska.ne.gov

C. Other Areas
   1. Tri-state corner (Map location 2, highest point in Nebraska,). This remote area can be reached by driving south from Bushnell (south 12.5 miles, west 4.2 miles, south 1 mile, west 2 miles, & south 2 miles, to nearly the Kansas line). Just past the H. Constable home there is a right turn with a cattle guard, and a sign reading "Panorama Point, Highest Point in Nebraska." After a mile there is another cattle guard, and another right turn (going north). It is then 0.3 miles to the high point. There is a charge ($4.00 in 2006) for driving to the high point). Look for lark buntings, horned larks,

McCown's longspurs and mountain plovers in the vicinity (Rosche, 1994).

   2. Lodgepole Creek (Map location 3). This nearly dry creek should attract warblers and other passerines.

8. Morrill County (Map 6)

Morrill County has nearly 30,000 acres of wooded habitats, some 660,000 acres of grasslands, and almost 5,000 acres of surface water. There are tourist accommodations at Bayard and Bridgport.

A. Federal Areas
   1. Chimney Rock National Historic Site. (Map location 1). Area 83 ac. Chimney Rock is located near Bayard, and is worth investigating for nesting golden eagles. An old cemetery lies to the northwest of Chimney Rock (see map), and burrowing owls are often found in a nearby prairie dog colony. Lazuli buntings are common in brushy areas.
www.nps.gov/chro

B. State Areas
   1. Bridgeport SRA (Map location 4). Area 128 acres. In Bridgeport. http://outdoornebraska.ne.gov/parks/guides/parksearch/showpark.asp?Area_No=35

   2. Chet and Jane Fleisbach (Facus Springs) WMA (Map location 2). This WMA preserves one of the best saline marshes in the North Platte Valley. It is a major stopover point for migrant shorebirds, and also attracts ducks during migration. Some shorebirds such as American avocets and Wilson's phalaropes also nest, and the cinnamon teal has also nested here (Rosche, 1994).

C. Other Areas
   1. Saline marsh near Bridgeport (Map location 3). Like Chet and Jane Fleishbach WMA, this saline marsh near Bridgeport attracts great numbers of shorebirds during migration. In wet years other marshes may occur here too. Accessible from the railroad track.

   2. Courthouse Rock and Jail Rock (Map location 6). These famous Oregon Trail landmarks have had nesting golden eagles (on Jail Rock), and breeding rock wrens are common. By driving west on state

route 88 to Redington one can take the Redington gap road to Facus Springs and Bridgeport, or go south from Redington (see below). www.nps.gov/oreg/planyourvisit/site6.htm

3. Redington Gap road (Map location 5) and road south of Redington (Map location 7). By driving south past Facus Springs one passes over a long, eroded line of hills (Redington Gap), and many western species typical of the high plains may be seen. Within a mile of turning south off US routes 26/92 one passes a meadow that supports a good population of savannah sparrows (rare in Nebraska). By continuing to Redington and going south from there for 4.5 miles, then take a left fork, and go 3 more miles until pines appear on a north-facing slope. This area (eastern end of Wildcat Hills) supports a good population of Cassin's kingbirds, plus the easternmost known pinyon jay population, and such western birds as western wood-pewees and common poorwills (Rosche, 1994).

9. Cheyenne County

Cheyenne County has almost no surface water, about 7,000 acres of wooded habitats, and over 210,000 acres of grasslands or farmlands. There are tourist accommodations at Lodgepole and Sidney.

A. Federal Areas: None

B. State Areas: None

10. Garden County (Maps 3, 7)

Garden County has over 22,000 acres of surface water, about 500 acres of wooded habitats, and nearly 900,000 acres of grasslands or farmlands. There are tourist accommodations at Oshkosh.

A. Federal Areas
1. Crescent Lake National Wildlife Refuge. (Map 3, location 1). 40,900 acres. This is one of the great wilderness refuges in America, and it supports a greater bird diversity than any other Nebraska site except the Lake McConaughy area (See checklist in supplement). However, it is about 30 miles from the nearest source of gas, food, or lodging, and one must plan accordingly, taking

a tow rope if possible, and never parking on bare sand. Rather, park or turn around on level, grassy meadows if possible. Water and a toilet are available at the refuge headquarters. Goose Lake near the headquarters is excellent for eared grebes, and both Crescent Lake and Smith Lake have good populations of western grebes (and some Clark's grebes). Rush Lake (just outside the refuge boundary) has breeding ruddy ducks, canvasbacks, redheads and black-crowned night herons. The area near Border Lake is best for avocets, black-necked stilts, cinnamon teal, Wilson's phalaropes and other shorebirds attracted to saline water conditions; Border Lake marks the boundary of such hypersaline conditions (see map 7). On most visits no other people will be seen, but the birding will be spectacular, and well worth the long ride over sand roads. A sharp-tailed grouse blind holds about three people; reservations for its use are required. Classified as a Nebraska Important Bird Area. For maps, a bird checklist or other information call the refuge office at 308/635-7851 or 308/762-4893. www.fws.gov/crescentlake

B. State Areas
1. Ash Hollow State Historical Park (Map 7, locations 1 & 3). This historically interesting park has a wide variety of habitats, from exposed rocky bluffs that are used by great horned owls, American kestrels, and sometimes prairie falcons, through grassy wet meadows where bobolinks and eastern meadowlarks are present, to riparian wooded habitats used by warbling vireos and other wooded habitats songbirds. There is also upland grassland, with blue grosbeaks and spotted towhees in shrubby areas, and scattered yuccas where field and grasshopper sparrows sometimes perch. An air-conditioned interpretive center provides welcome relief from oppressive summer temperatures. The nearby U.S. 26 bridge across the North Platte provides views of many marshland species, including least bitterns on rare occasions. A one-mile trail leads from the parking lot off U.S. 26 to Windlass Hill, where ancient pioneer wagon ruts are still easily visible. http://outdoornebraska.ne.gov/parks/guides/parksearch/showpark.asp?Area_No=8

2. Clear Creek Waterfowl Management Area (Map 7, location 2). See Keith County account.

C. Other Areas

1. Oshkosh Sewage Lagoons (Map 7, location 7). These lagoons are reached by driving south on Route 27 for 0.5 miles from Oshkosh, turning east, and driving until the lagoons appear on the south side of the road. Three lagoons are accessible by walking. They attract a surprising array of waterfowl, including breeding wood ducks and even nesting ruddy ducks (Rosche, 1994).

11. Deuel County (Map 7)

Deuel County has more than 1,000 acres each of wooded habitats and surface water, and nearly 64,000 acres of grasslands or farmlands. There are tourist accommodations at Big Springs and Chappell.

A. Federal Areas: None

B. State Areas

1. Bittersweet WMA. (Map location 6). Area 76 acres. Consists of South Platte River frontage.

2. Goldeneye WMA. (Map location 5). Area 25 acres. Includes 11 acres of wetland, with a large prairie dog town beside it, and associated short-grass prairie birds, including horned larks, Lapland longspurs, and ferruginous hawks. Access is from Big Springs I-80 exit, go 1 mile south, 3 miles west, and then back over I-80.

3. Goldenrod WMA. (Map location 4). Area 97 acres, all mixed-grass prairie and woods habitat.

McCown's Longspur

The West-Central Region: Sandhills Country

This portion of the state includes two of the very best bird-finding localities in the state, namely Valentine and Fort Niobrara National Wildlife Refuges. These two locations have bird lists that are among the largest in the state (see supplement checklists). Additionally, it includes those parts of the Niobrara and Platte Valleys that lie in the middle of the transition zone between the Rocky Mountain coniferous forest and eastern deciduous forest biogeographic regions. These transition or "suture zones" include areas of hybridization between several species or nascent species pairs of birds that are now in secondary contact, after having been isolated geographically for much or all of the Pleistocene geologic period. This transition zone is very wide in the Platte Valley, but is compressed to a distance of less than 100 miles in the Niobrara Valley, most of which is now included within the boundaries of the Fort Niobrara National Wildlife Refuge and the Nature Conservancy's wonderful Niobrara Valley Preserve. The Niobrara Valley also support breeding populations of several eastern wooded habitats species that are otherwise mostly limited to Nebraska's Missouri Valley, including the wood thrush, black-and-white warbler, American redstart, ovenbird and scarlet tanager. Along the upper Niobrara Valley several western or northern bird species likewise extend eastwardly, including the common poorwill, red-breasted nuthatch, chestnut-collared longspur, red crossbill, western wood-pewee, spotted towhee, black-headed grosbeak, lazuli bunting and Bullock's oriole. All but the first four of these might hybridize with eastern relatives along this valley corridor.

Beyond these regions, and lying directly between them, the region mostly consists of the Nebraska Sandhills. This region represents the largest natural ecosystem in the state, covering nearly 19,000 square miles, or almost a quarter of the state. It is also the largest remaining grassland ecosystem in the country that is still virtually intact both faunistically and floristically. It is a land with far fewer people than cattle, where the roads are few and where tourist facilities and accommodations are almost non-existent. Those roads that do exist are little-traveled, and often consist of only slightly improved sandy trails leading to ranches. But the region is filled with breathtaking vistas, spectacular bird populations in the hundreds of lakes and marshes, and a pioneer spirit that requires everyone to help his neighbor, or indeed any stranger who happens to fall afoul of trouble while on the road. It is a land designed for naturalists who would like to study virtually unaltered prairie ecosystems, and who are prepared to deal with nature on its own terms. Many waterbirds and shorebirds nest almost only here in Nebraska, such as the American wigeon, canvasback, redhead, ruddy duck, Wilson's snipe, Forster's tern, marsh wren and swamp sparrow. A summary of the natural history of the Nebraska Sandhills, along with an annotated bird checklist (277 species) may be found in *This Fragile Land: A Natural History of the Nebraska Sandhills,* by P. A. Johnsgard (see references for complete citation).

1. Cherry County (Map 8)

Cherry County is by far the largest county in the state, and mostly consists of Sandhills habitat, with 3.7 million acres of grasslands, about 17,000 acres of wooded habitats, and 41,000 acres of surface wetlands. There are tourist accommodations at Merriman and Valentine.

A. Federal Areas
1. Fort Niobrara National Wildlife Refuge (Map location 1). Area 19,122 acres. This refuge, originally established to protect bison and other large game animals, lies on the western edge of the east-west ecological transition zone between forest

types, and thus has a fine mixture of eastern and western avifauna. Western-eastern species pairs that occur and may hybridize include such forms as western and eastern wood-pewees, black-headed and rose-breasted grosbeaks, eastern and spotted towhees, and Bullock's and Baltimore orioles. A checklist of the refuge's bird species (excepting accidentals) may be found in the supplement. A total of 201 species (76 breeders) have been reported here. About two-thirds of the refuge consists of Sandhills prairie, and the rest is mostly of mixed riparian hardwoods. There is a good population of sharp-tailed grouse, and wild turkey viewing blinds are available. There are also breeding burrowing owls, yellow-breasted chats, American redstarts, grasshopper and savannah sparrows, and both meadowlarks. The refuge address is Hidden Timber Star Rte., Valentine, NE, 69201 (402/376-3789), www.fws.gov/fortniobrara. The John & Louise Seier National Wildlife Refuge (2,400 acres) in Brown County is 25 miles south of Bassett, is managed through Fort Niobrara. It is under development and not yet open to the public.

2. Valentine National Wildlife Refuge (Map location 7). Area 71,516 acres. This is Nebraska's largest national wildlife refuge, and one that rivals Crescent Lake in its bird diversity, with 221 species (93 breeders) reported. A checklist (excluding accidentals) is included in the supplement. Most of the refuge consists of Sandhills prairie, with dunes 40 to 200 feet high, and intervening interdune depressions that often contain shallow, marshy lakes. Some of the lakes are open for canoeing or boating, offering great birding opportunities. Driving on the sandy trails requires care; a supply of water and a tow rope are recommended. Several prairie-chicken and sharp-tailed grouse leks are present in the refuge, and two public-use blinds are on prairie-chicken and sharp-tail leks. Wetlands offer breeding habitat for eared, western and pied-billed grebes, a dozen species of waterfowl, and shorebirds such as soras, Wilson's snipes, and American avocets. The higher grasslands offer views of long-billed curlews, upland sandpipers, and Swainson's hawks. Classified as a Nebraska Important Bird Area. The refuge manager's address is the same as that of the Fort Niobrara refuge (402/376-3789).
www.fws.gov/valentine

3. Samuel R. McKelvie Unit, Nebraska National Forest (Map location 2). Area 115,703 acres. This section of forest is similar to that of the Bessey Ranger District (see ThomasCounty) but is not so rich in migrants. The adjoining Merritt reservoir usually has manywater birds, especially during migration. There are sharp-tailed grouse leks on the area, but one must provide one's own blind. For information phone 308/533-2257.
www.fs.fed.us/r2/nebraska/units/mckelvie/mckelvienf.html

B. State Areas

1. Schlegel Creek WMA. (Map location 3). Area 600 acres. Consists of Sandhills grassland including two miles of Schlegel Creek. No facilities for camping are present.

2. Big Alkali WMA. (Map location 5). Area 842 acres. Consists of 47 lakeside acres plus an 842-acre Sandhills lake. Campground present.

3. Ballard's Marsh WMA. (Map location 6). Area 1,561 acres. Includes a large marsh and adjoining Sandhills grasslands or farmlands. Campground present.

4. Smith Falls State Park. Located 3 miles west and 4 miles south of Sparks. Area 244 acres. The river can be crossed by a new walking bridge for a view of the falls.
http://outdoornebraska.ne.gov/parks/guides/parksearch/showpark.asp?Area_No=308

5. Merritt Reservoir WMA (Map location 4). Located 26 miles southwest of Valentine. Area, 2,906 acres, reservoir, 350 acres upland Sandhills. This area abuts National Forest land to the north. The reservoir attracts migrant waterfowl, pelicans, western grebes, and other species.

6. Cottonwood/Steverson Lakes WMA. This WMA has 2,919 acres, encompassing three lakes (Cottonwood, Steverson and Home Valley). The western end of Steverson Lake has a fen, and associated cold-climate plants that are relicts of the Pleistocene period. Other fens also occur in this headwaters area of the North Loup River. Located south about 30 miles of Cottonwood Lake SRA on State Highway 61.

7. Cottonwood Lake SRA. Area 180 land acres, 60-acre lake. Located 1 mile southeast of Merriman. Camping facilities present. Canada geese breed here. Five miles east of Merriman is a marsh where trumpeter swans have nested, off the north side of US highway 20.
http://outdoornebraska.ne.gov/parks/guides/park-search/showpark.asp?Area_No=53

8. Bowring Ranch State Historical Park. Located 1 mile north of Merriman. Trumpeter swans forage on a marsh just north of this park, a working cattle ranch. Park entry permit required.
www.byways.org/explore/byways/16471/places/38093

2. Keya Paha County (Map 9)

Keya Paha County has over 37,000 acres of wooded habitats, over 400,000 acres of grasslands, and about 1,300 acres of surface water. The only tourist accommodations are in Springview.

A. Federal Areas: None

B. State Areas
1. Cub Creek Recreation Area. (Map location 2). Area 300 acres.

2. Thomas Creek WMA. (Map location 3). Area 692 acres. Steep topography around Thomas Creek, with grassland on the hills and wooded creek-bottom.

C. Other Areas
1. Niobrara Valley Preserve. See Holt County.
www.nature.org/wherewework/northamerica/states/nebraska/preserves/art24963.html

3. Brown County (Map 10)

Brown County has over 21,000 acres of wooded habitats, nearly 700,000 acres of grasslands, and about 8,000 acres of surface water. There are some tourist accommodations at Ainsworth.

A. Federal Areas: None

B. State Areas
1. Bobcat WMA (Map location 3). Area 893 acres.

Nearly 90 percent of this area consists of steep pine- and cedar-covered canyons. Plum Creek passes through. The remainder is Sandhills grassland.

2. School Land WPA & Keller Park SRA (Map location 4). Area 836 acres. (WMA 640 acres, SRA 196 acres). These areas consist of native prairie, wooded canyons, Bone Creek, and five small fishing ponds stocked with trout and other game fish. The ponds attract ducks, eagles, and other water birds, the prairies support grassland sparrows, and the mixed wooded habitats have a variety of both coniferous and deciduous forest birds including wild turkeys, scarlet tanagers and American redstarts.
http://outdoornebraska.ne.gov/parks/guides/park-search/showpark.asp?Area_No=250

3. Pine Glen WMA (Map location 5). Located 7 miles west and 6.5 miles north of Bassett. It consists of 960 acres of canyons, a trout stream, and mixed grasslands and wooded habitats. No facilities.

4. Long Pine WMA. (Map location 6). Area 160 acres. Consists of about 85 percent pine and red cedar wooded habitats, and the rest is native Sandhills grassland, bisected by Long Pine Creek. The terrain is steep and camping facilities are primitive. Located just off Highway 20 near the town of Long Pine.

5. South Twin Lake WMA. (Map location 7). Area 160 acres. Consists of a 60-acre lake and Sandhills grassland. This and the next three WMAs are similar, and should offer birders some excellent views of Sandhills wildlife.

6. American Game Marsh WMA (Map location 8). Area 160 acres. Consists of a large Sandhills marsh and surrounding grassland. No facilities.

7. Long Lake WMA (Map location 9). Area 30 acres upland, 50 acres Sandhills lake.

8. Willow Lake WMA (Map location 10). Area 511 acres. A Sandhills lake and surrounding grassland.

C. Other Areas
1. Niobrara Valley Preserve (Map location 1; headquarters at location 2). Area circa 56,000 acres

including about 25 miles of the Niobrara River, in the heart of the transition zone between western coniferous and eastern deciduous forest types, There is a bird checklist of 186 species, including a list of 75 definite and 30 more possibly breeding species (Brogie & Mossman, 1983). The Preserve has been identified as a Globally Important Bird Area by the American Bird Conservancy. Among the breeding birds of special interest are the eastern and western forms that hybridize here, such as the Baltimore and Bullock's orioles, the lazuli and indigo buntings, and the rose-breasted and black-headed grosbeaks. The eastern and western wood-pewees both also occur here. Two trails radiate out from the headquarters that pass through several forest types and the Sandhills prairie vegetation on the uplands. Each trail has a short loop and a long loop; the northern one is somewhat longer (three miles) and steeper. For information phone 402/722-4440. The preserve lies within the Niobrara National Scenic River District, which extends for 76 miles and is a popular canoeing destination (call 402/376-3241 for information.)
www.nature.org/wherewework/northamerica/
states/nebraska/preserves/art24963.html

4. Rock County

Rock County is, despite its name, mostly comprised of Sandhills habitat, with about 11,000 acres of wooded habitats, over 600,00 acres of grasslands, and about 11,000 acres of surface water. The only tourist accommodations are at Bassett.

A. Federal Areas
1. John & Louise Seier National Wildlife Refuge. Area 2,400 acres. Located in southwestern Brown County 25 miles south of Bassett (via U.S. Highway 183) and several miles west (via county road). Sandhills grasslands and marshes, including Hornburger Lake andthe headwaters of Skull and Bloody creeks. This refuge is still under development and not yet open to the public. It is administered by the Fort Niobrara National Wildlife Refuge, which should be contacted for information.

B. State Areas
1. Twin Lakes WMA. Located 18 miles south and 2 miles east of Bassett. It includes 113 acres of surface water (two lakes) and 30 acres of grassland.

C. Other Areas
1. Hutton Niobrara Ranch Wildlife Sanctuary. 4,919 acres. Located 15 miles north of Bassett (via State Highway 7), on bottomlands and uplands south of the Niobrara River. This former ranch was given to the Audubon Society of Kansas by its prior owner, Harold Hutton. It will be used as a sanctuary and will be managed both for cattle and wildlife. An adjacent 160-acre farm with a four-bedroom home will be used by students, volunteers and other visitors. Red cedars that have invaded the grasslands are being removed to restore native prairie, and a cultivated field will be reseeded to native grasses. Wet meadows along the river will be managed to attract bobolinks and other meadow-nesting birds. There are also plans to introduce prairie dogs. An opening date has not yet been announced.

5. Grant County

Grant County is a sparsely populated Sandhills county with only about 400 acres of wooded habitats, about 3,500 acres of surface water, and over 460,000 acres of grassland. There are no tourist accommodations in the county.

A. Federal Areas: None

B. State Areas: None

6. Hooker County

Hooker County is another sparsely populated Sandhills county, with 1,800 acres of wooded habitats, about 400 acres of surface water, and over 450,000 acres of grasslands or farmlands. There are tourist accommodations at Mullen. A motel (Sandhills Motel) owner (Mitch Glidden) there offers canoe rentals and operates spring sunrise trips to prairie-chicken and sharp-tailed grouse leks, where birders can view displays of both species from the comfort of a school bus. Phone 888/278-6167 (or www.hooker-county.com/sandhillsmotel.html).

A. Federal Areas: None

B. State Areas: None

7. Thomas County (Map 11)

Thomas County is in the heart of the Sandhills but has over 16,000 acres of wooded habitats, most of which lies in the planted pine "forest" near Halsey. There are about 1,500 acres of surface water, (mostly Middle Loup and Dismal rivers), and almost 380,000 acres of grassland. There are tourist accommodations at Halsey and Thedford.

A. Federal Areas
1. Bessey Ranger District, Nebraska National Forest (Map location 1; locations 2 and 3 indicate traditional lek locations, with public-use blinds holding up to 4 people. Area 90,445 acres. Grasslands around and in this planted "forest" support greater prairie-chickens, sharp-tailed grouse, upland sandpipers, horned larks, and western meadowlarks. The conifers provide habitat for great horned owls, black-capped chickadees, and red crossbills. Brushy and riparian thicket areas attract several woodpeckers, brown thrashers, towhees, chipping sparrows and Baltimore orioles. At least six warbler species nest here, including yellow, black-and-white, American redstart, ovenbird, common yellowthroat, and yellow-breasted chat. Three vireos (Bell's, warbling and red-eyed) also nest here. Townsend's solitaires and red-breasted nuthatches are common in winter. There is a bird checklist available at the headquarters, where information on the grouse blinds is also available (Boyle & Bauer, 1994). One small prairie dog town is still present, but regrettably one other has been destroyed by hunters. A fire in the 1960s burned much of the forest, as did another in 2006 but most of the original 25,000 acres of woods still survive. Three grouse viewing blinds exist, each holding about four adults. There is a 6-mile hiking trail that begins at the parking lot off State Highway 2. For information contact the Forest Service office at PO Box 38, Halsey, NE 691142 (308/533-2257).
www.fs.fed.us/r2/nebraska/units/brd/brd.html

B. State Areas: None

8. Blaine County (Map 11)

Blaine County is a Sandhills county with 1,600 acres of wooded habitats, about 1,000 acres of surface water, and nearly 440,000 acres of grassland.

There are no tourist accommodations in the county.

A. Federal Areas: None

B. State Areas
1. Bessey Unit, Nebraska National Forest (Map location 1). See also Thomas County. www.fs.fed.us/r2/nebraska/units/brd/brd.html

2. Milburn Dam WMA. Located 14 miles southeast of Brewster. Consists of 672 acres of Middle Loup River Valley, with extensive mud flats present around the reservoir.

9. Loup County

Loup County is a Sandhills County with about 7,000 acres of surface water, nearly 5,000 acres of wooded habitats, and over 325,000 acres of grasslands or farmlands. The only tourist accommodations are at Taylor.

A. Federal Areas: None

B. State Areas
1. Calamus Reservoir SRA/WMA. Area 10,312 acres total; reservoir 5,124 acres. Located near county boundary; see Blaine County.
http://outdoornebraska.ne.gov/parks/guides/park-search/showpark.asp?Area_No=275

10. Arthur County (Map 12)

Arthur County is another sparsely populated Sandhills county with almost 3,000 acres of surface water, about 200 acres of wooded habitats, and 427,000 acres of grasslands or farmlands. There are no tourist accommodations in the county.

A. Federal Areas: None

B. State Areas: None

C. Other Areas
1. Marshes near McPherson County border (Locations of better marshes shown by arrowheads). These Sandhills marshes and creeks are often used by trumpeter swans.

11. McPherson County

McPherson County is a very sparsely populated Sandhills county, with about 600 acres of surface water, 1,000 acres of wooded habitats, and over 520,000 acres of grasslands or farmlands. There are no tourist accommodations in the county.

A. Federal Areas: None

B. State Areas: None

C. Other Areas
1. Several large marshes occur near the Arthur County border, such as Diamond Bar Lake about 3.5 miles south of State Highway 92 (the turning about 5 miles east of the Arthur County line) on a secondary road, and with only a sandy trail for close access. Trumpeter swans are regular here. White Water Lake, Dry Lake and Brown Lake are located north of State Highway 92, and are also on unmarked secondary roads. See also Arthur County.

12. Logan County

Logan County is a Sandhills county with only 250 acres of surface water, about 300 acres of wooded habitats, and over 520,000 acres of grasslands or farmlands. No tourist accommodations are present in the county.

A. Federal Areas: None

B. State Areas: None

13. Custer County

Custer County is a mostly Sandhills county with about 2,500 acres of surface water, over 10,000 acres of wooded habitats, and about 1.1 million acres of grasslands or farmlands. There are tourist accommodations at Arnold, Broken Bow, Callaway and Sargent.

A. Federal Areas: None

B. State Areas
1. Victoria Springs SRA. Area 60 acres. Small campground and lake surrounded by deciduous forest with cottonwoods. State park entry permit required.
http://outdoornebraska.ne.gov/parks/guides/parksearch/showpark.asp?Area_No=179

2. Pressey WMA. Area 1,640 acres. Located 5 miles north of Oconto. Consists of South Loup Valley lands, hills and steep canyons mostly covered by grasslands or farmlands. Toilets and a campground are present, as are hiking trails. There are sharp-tailed grouse on the area, as well a great blue heron rookery.

3. Arcadia Diversion Dam SRA. Area 925 acres. This area of the Middle Loup River Valley consists mostly of grasslands and tree plantings, but with deciduous wooded habitats lining the river. There are some campgrounds on both sides of the river. Located 8.5 miles northwest of Arcadia.

14. Keith County (Map 13)

Keith County is notable in having over 37,000 acres of surface water, nearly 6,000 acres of wooded habitats, and over 420,000 acres of mainly Sandhills grasslands or farmlands. There are tourist accommodations at Keystone, Lemoyne, Ogallala and Paxton.

A. Federal Areas: None

B. State Areas
1. Clear Creek WMA. (Map location 1). Area 5,709 acres. Partly developed as Clear Creek Refuge (2,500 acres, west half), and also as a controlled hunting area. The latter includes the west end of Lake McConaughy and the Platte River inflow area. The low meadows support nesting bobolinks and probably breeding Wilson's snipes, and the tall tree groves hold many breeding passerines. White pelicans are common, and least bitterns have been sighted. One of the state's best birding areas, but mosquitoes can be a problem during summer. Barn owl nest cavities usually can be seen in the cutbanks at the turnoff from the main highway; nests in this part of the state are usually in such excavated sites rather than in old buildings. Rosche (1994) has described this area and its birds very well, which is the state's only known nesting area for Clark's grebe.

2. Lake McConaughy SRA (Map location 2). Area 6,492 acres. The SRA occupies much of the north side of this reservoir, the largest body of water in Nebraska. A small area on the south side is also included (Map location 6). This area has the largest bird list of any location in the state, including about 340 species, with 104 known breeders, 17 additional possible breeders, and about 200 transients (Brown et al., 1996; Brown & Brown, 2000). The large water area attracts vast numbers of migrant waterfowl, grebes (especially western grebes), gulls (including many rarities) and shorebirds (Rosche, 1994). A good spotting scope is needed to cover this vast reservoir, but many of the waterfowl congregate near the spillway during winter, or (in the summer) toward the western end of the lake (see Clear Creek WMA account). Large numbers of bald eagles also build up in winter, attracted by dead fish and the wintering duck and goose populations. The checklist for the North Platte Valley provided in the supplement is largely based on the Lake McConaughy checklists by Rosche (1994) and Brown et al. (1996). Well over 100 miles of shoreline are present along the lake, with the southern shoreline rocky and steep, and the northern shore sandy. These support nesting piping plovers and least terns. Classified as a Nebraska Important Bird Area. Some of the rarer birds found here are trumpeter swan, cinnamon teal, Clark's grebe, all three jaegers, Sabine's gull, and common tern. Snowy plovers have nested in recent years, as do piping plover and least tern. Both eastern and western wood-pewees occur, as do east-west species pairs of orioles, grosbeaks and buntings. For information phone 866/386-2862. http://outdoornebraska.ne.gov/parks/guides/park-search/showpark.asp?Area_No=99

3. Kingsley Dam and Lake Ogallala SRA (Map location 3). Area 339 acres. Kingsley Dam offers a good vantage point for birds both on the deeper end of Lake McConaughy and on the shallower and much smaller Lake Ogallala located at the base of the dam. Lake Ogallala (and its eastern end, often called Lake Keystone) receives the spillway water from Lake McConaughy, and its level fluctuates greatly. However, it is very attractive to migrant ducks, ospreys, Caspian terns, cliff swallows, gulls American white pelicans, double-crested cormorants and other summering species,

and is used by Canada geese and by numerous bald eagles in winter. An eagle-watching blind is available during peak periods, when 200-300 eagles are sometimes present. It is available from late December through early March. Thursdays & Fridays, 8 a.m. to noon, Saturdays and Sundays, 8 a.m. to 4 p.m. The northern shorelines of Lake Ogallala has deciduous wooded habitats with a rich array of nesting passerines, but lake fluctuations limit nesting for aquatic species. Lake Ogallala SRA is classified as a Nebraska Important Bird Area. Some of the rare gulls seen here include mew, Thayer's, glaucous and lesser black-backed. Rare waterfowl include trumpeter swan, greater scaup, all scoters, long-tailed duck and Barrow's goldeneye. For eagle-viewing information phone 308/284-2332.

4. Cedar Point Biological Station (Map location 4). Although an extension of the University of Nebraska and a summer field station, and thus not usually open to the public, ornithological research here has made its avifauna the best-known of any area in the state (Brown & Brown, 2000). Ornithology courses have been taught here on a regular basis since 1977, and studies on species such as the cliff swallow and orchard oriole have been of national significance. Classified as a Nebraska Important Bird Area.

5. Ogallala Strip WMA (Map location 5). Area 453 acres, includes 2.5 miles of river frontage. This stretch of riparian wooded habitats supports many of the same species found around Lake Ogallala, such as house wren, yellow warbler, common yellowthroat, eastern and western kingbirds, killdeer, and others. Mississippi kites now breed in nearby Ogallala.

6. Lakeview campground (Map location 6). The road leading down the canyon to Lakeview, and a similar road leading to Eagle Canyon 6 miles farther west, may offer views of rock wrens, turkey vultures, rough-winged swallows and, with luck, occasional prairie falcons or ferruginous hawks. Turkey vultures nest along the south side of the reservoir, usually in eroded crevices or recesses well out of view. These roads are often in poor condition, and caution must be exercised when driving over them. For information phone 866/386-2862.

15. Perkins County

Perkins County is a high plains county with only about 200 acres of surface water, about 1,000 acres of wooded habitats, and 125,000 acres of grasslands or farmlands. There are no tourist accommodations. However, in wet springs hundreds of shallow playa lakes develop in northern Perkins and southern Keith counties, attracting large numbers of waterfowl and shorebirds. The state's largest remaining area of sandsage is southeast of Grant, near the Chase County line

A. Federal Areas: None

B. State Areas: None

16. Lincoln County (Map 14)

Lincoln County straddles the Platte Valley with nearly 10,000 acres of surface water, about 36,000 acres of wooded habitats, and about 1.2 million acres of grasslands or farmlands. There are tourist accommodations at North Platte and Sutherland.

A. Federal Areas: None

B. State Areas
1. Sutherland Reservoir SRA. (Map location 1). Area 3,020 acres. reservoir, 37 acres upland. Rosche (1994) refers to this site as "the gull capitol of western Nebraska," with ten species having been observed. These include such rarities as Thayer's, glaucous, great and lesser black-backed, and even Ross' gull. There are often large flocks of wintering grebes, diving ducks, double-crested cormorants, and American white pelicans during mild winters. During spring large flocks of snow, greater white-fronted and occasional Ross' geese stop here. http://outdoornebraska.ne.gov/parks/guides/park-search/showpark.asp?Area_No=171

2. Malony Reservoir SRA (Map location 2). Area 1,600 acres. reservoir, 1,732 acres upland. This lake is used during spring by American white pelicans and double-crested cormorants, and many shore-birds when the water levels subside (Pettingill, 1981).
http://outdoornebraska.ne.gov/parks/guides/park-search/showpark.asp?Area_No=112

3. Jeffrey Canyon WMA and Reservoir. (Map location 3). Area 900 acres reservoir, 35 acres upland. This area consists of canyon-and-upland topography, with grasses and scattered deciduous trees and cedars. Very limited public access (at the dam and boat ramp).

4. North River Wildlife WMA (Map location 4). Area 681 acres, 2 miles of river frontage. There are woods along the river, and grassland beyond that is used by sandhill cranes. This is one of the westernmost crane roosting sites; the birds use the southeastern part of the area, in less-than-ideal roosting habitat. However, recent habitat improvements might make the conditions more suitable for cranes.

5. Muskrat Run WMA. (Map location 5), Area 224 acres. Mostly riparian wooded habitats and marshy areas.

6. East Sutherland WMA. (Map location 6). Area 27 acres upland, 8 acres lake.

7. Hershey WMA. (Map location 7). Area 53 acres upland, 80 acres lake.

8. East Hershey WMA. (Map location 8). Area 20 acres upland, 20 acres lake.

9. Birdwood Lake WMA. (Map location 9). Area 20 acres upland, 13 acres lake.

10. Fremont Slough WMA. (Map location 10). Area 30 acres upland, 11 acres lake.

11. Platte WMA. (Map location 11). Area 242 acres upland, 0.5 miles of river frontage. Mostly riparian wooded habitats.

12. Ft. McPherson Cemetery (Map location 12). Area 30 acres, with pond.

13. West Brady WMA (Map location 13). Area 10 acres upland, 6 acres lake.

14. Chester Island WMA (Map location 14). Area 69 acres, ponds. Includes 0.3 miles of river frontage.

15. Box Elder Canyon WMA. Not shown, located 3 miles south and 2.5 miles west of Maxwell. This 20-acre site consists of native grasslands and deciduous wooded habitats along the Tri-County Supply Canal. Nearby are Cottonwood and Snell canyons, both on private lands, but all three canyons support black-headed grosbeaks, Say's phoebes and rock wrens in summer, and mountain bluebirds, eastern bluebirds, Townsend's solitaires and cedar waxwings during winter.

16. Wellfleet WMA. Not shown, this area is just west of the village of Wellfleet, or 20 miles south of North Platte. Comprising only 65 acres along Medicine Creek, it provides a diversity of habitats that usually attracts a wide variety of small passerines and water birds (Rosche, 1994). In the summer yellow-breasted chats, Bell's vireos and other bush-loving species are common.

17. Wapiti WMA. 1,280 acres. Loess hills, mixed grass prairie, trees and shrubs, and about 40-50% cedar woodlands. Species include rock wrens, Say's phoebes, black-headed grosbeaks and some eastern species such as great crested flycatchers. From I-80 interchange at Maxwell drive south about 2 miles, turn south at Y junction onto Cottonwood Road, go 7 miles; stay at right at Y junction to reach Effenbeck Road, then after 1.5 miles turn right onto minimum maintenance road for about one more mile. Road is hazardous when wet.

C. Other Areas
1. North Platte Sewage lagoons (Map location 15). These sites are reached by leaving I-80 at exit 179, and going north on spur road L56G. Cross the South Platte River and turn east on a dead-end gravel road that will take you to the lagoons. These lagoons attract many water birds during migration. Pettingill (1981) described several wetlands to the north of North Platte (Whitehorse Marsh, Jackson Lake, and Ambler Lake) that support typical Sandhills marsh birds and waterfowl.

2. Birdwood Creek. Go north out of Sutherland on N. Prairie Trace Road (off Hwy 30, east side of town), cross the N. Platte River, and turn right on gravel road. Follow this road until you cross Birdwood Creek, then turn north and follow the creek for 6-7 miles. Look for trumpeter swans and other waterfowl.

17. Dawson County (Map 15)

Dawson County is another Platte Valley county with about 8,000 acres of surface water, over 17,000 acres of wooded habitats, and over 250,000 acres of grasslands.

A. Federal Areas: None

B. State Areas
1. Willow Island WMA (Map location 1). Area 45 acres upland, 35-acre lake, & riparian wooded habitats.

2. East Willow Island WMA (Map location 2). Area 16 acres upland, 21 acres wetland. Includes 0.3 miles of river frontage; mostly riparian wooded habitats.

3. West Cozad WMA (Map location 3). Area 19 acres upland, 29-acre lake.

4. Cozad WMA (Map location 4). Area 182 acres upland, 16 acres wetland, 0.5 miles of river frontage.

5. East Cozad WMA (Map location 5). Area 18 acres; all upland.

6. Darr Strip WMA (Map location 6). Area 976 acres; 767 acres land, 2.5 miles of river frontage.

7. Dogwood WMA (Map location 7). Area 402 acres, 10 acres lake, 1.5 miles of river frontage.

8. Midway Lake WMA (Map location 8). A reservoir near the Tri-County Canal; at its upper (southern) end is Midway Canyon, an eroded area of loess hills.

9. Gallagher Canyon. (Map location 9). Area 400 acres reservoir, 424 acres upland. Another canyon in the loess hills and associated reservoir.

10. Plum Creek WMA (Map location 10). Area 152 acres, 320 acres reservoir.

11. Johnson Lake SRA (Map location 11). Area 2,061 acres reservoir, 81 acres upland. Located in Gosper County, 7 miles south of Lexington on U.S. HWY 283, just south of the Dawson and Gosper

County line. Habitat includes a 2,000-acre reservoir and 68 acres of upland. The best birding is in late fall, winter and early spring when there are fewer people around. It is a good area for gulls, waterfowl, cormorants, eagles, loons and grebes. Excellent camping and water sports are hallmarks of this ever-popular area. Though relatively small at 68 acres, the SRA packs a lot of outdoor fun and provides three access points to the lake. It's at the heart of a complex of lakes on the Tri-county canal. Camping, picnicking, swimming and an office on site. (Courtesy Eric Volden) Elwood Reservoir (1,330 acres) is nearby.
http://outdoornebraska.ne.gov/parks/guides/park-search/showpark.asp?Area_No=94

12. Bittern's Call WMA. Located about 10 miles north of Lexington on Highway 21. Consists of 80 acres of mixed upland and wetland habitat.

18. Chase County (Map 16)

Chase County is a high plains county with about 2,200 acres of surface water (nearly all reservoirs), about 1,400 acres of wooded habitats, and 290,000 acres of grasslands or farmlands. The only tourist accommodations are at Imperial. Some sandsage still exists in western Chase County, especially from Lamar south to the Dundy County border, and in northern Chase County

A. Federal Areas: None

B. State Areas
1. Enders Reservoir SRA (Map location 1). Area 3,643 acres upland, reservoir 2,146 acres. Nearly all open grassland, with rolling to rugged topography. Developed facilities. This large reservoir attracts large numbers of mallards and Canada goose; most of the western half of the reservoir and surrounding land is a wildlife refuge. The north side of the reservoir is sandsage prairie, probably supporting Cassin's sparrow. Two prairie dog towns totaling about 35 acres are present.
http://outdoornebraska.ne.gov/parks/guides/park-search/showpark.asp?Area_No=71

2. Enders Reservoir WMA (Map location 2). This area to the west of the reservoir is managed for big game and upland game hunting.

3. Wannamaker WMA (Map location 3). This area of 160 acres, located about one mile west of Imperial, mostly consists of planted grasslands and shelterbelts.

4. Champion Lake SRA (Map location 4). Area 13 acres. Park entry permit required.
http://outdoornebraska.ne.gov/parks/guides/park-search/showpark.asp?Area_No=44

5. Imperial Light Dam & Reservoir Pond. 640 acres. Municipal dam and reservoir. Sandsage and shortgrass prairie with a cattail marsh and riparian cottonwoods. Four miles south and two miles west of Imperial. Not shown on map.

19. Hayes County

Hayes County is a high plains county with less than 800 acres of surface water, about 2,000 acres of wooded habitats, and 255,000 acres of grasslands or farmlands. The only tourist accommodations are at Benkelman.

A. Federal Areas: None

B. State Areas
1. Hayes Center WMA. Area 78 acres. Located 12 miles northeast of Hayes Center. It consists of native high plains grasslands, scattered wooded habitats, and a 40-acre reservoir. The shrubby riparian vegetation attracts many passerines, and some eastern species such as eastern phoebes, red-bellied woodpeckers and northern bobwhites breed here. To the north, along State Highway 25, western birds such as Say's phoebe, rock wren and ferruginous hawk may at times be seen. A cattail marsh at the upper end of the reservoir supports green herons, Virginia rails and marsh wrens.

20. Frontier County (Map 17)

Frontier County is a high plains county with about 3,500 acres of surface water (nearly all reservoirs), 1,300 acres of wooded habitats, and almost 330,000 acres of grasslands or farmlands. There are no tourist accommodations.

A. Federal Areas: None

B. State Areas

1. Red Willow Reservoir SRA/WMA (Map location 1). Area 1,628 acres reservoir, 4,320 acres upland. Modern camping facilities are present. This reservoir in a water-poor region attracts good numbers of migratory water birds, including many geese and ducks. Burrowing owls should be searched for in the prairie dog town north of the county road near the Spring Creek arm, and at Prairie Dog Point on the Red Willow arm. Park entry permit required for the SRA. See also Red Willow County.
http://outdoornebraska.ne.gov/parks/guides/parksearch/showpark.asp?Area_No=149

2. Medicine Creek Reservoir & Medicine Creek SRA/WMA (Map location 2). Area of WMA 6,726 acres, SRA area 1,768 acres reservoir & 1,200 acres upland. There are 17 hiking trails present and both primitive and modern camping facilities exist. Park entry permit required for the SRA.
http://outdoornebraska.ne.gov/parks/guides/parksearch/showpark.asp?Area_No=113

21. Gosper County (Map 18)

Gosper County is a mostly high plains county with less than 4,000 acres of surface water, nearly 1,700 acres of wooded habitats, and nearly 140,000 acres of grassland. The only tourist accommodations are at Elwood.

A. Federal Areas

1. Victor Lake Waterfowl Production Area (Map location 2). Area 174 acres wetland, 64 acres upland.

2. Elley Lagoon Waterfowl Production Area (Map location 6). Area 33 acres wetland, 29 acres upland.

3. Peterson Basin Waterfowl Production Area (Map location 7). Area 527 acres wetland, 627 acres upland.

B. State Areas

1. Johnson Lake SRA (Map location 1). Area 2,061 acres reservoir, 81 acres upland. See also Dawson County.
http://outdoornebraska.ne.gov/parks/guides/parksearch/showpark.asp?Area_No=94

2. Elwood Reservoir WMA. Located 2 miles north of Elwood. Consists of a 1,330-acre reservoir and 900 adjacent acres of grassland and some wooded sites. No camping facilities.

3. J-2 Hydro Power Plant (not shown on map). Located 6 miles south and ½ mile east of Lexington, and operated by Central Nebraska Public Power District. Open seasonally for eagle-viewing. During winter months up to 100 bald eagles may gather along the power plant spillway, feeding on dead and stunned fish that have passed through the spillway. Viewing is from windows in the plant. Free access, but public viewing hours and days are limited. For information, call 308/995-8601.

22. Phelps County (Map 18)

Phelps County is a Platte Valley county at the western edge of the Rainwater Basin, with under 200 acres of permanent surface water (plus temporary wetlands), 3,800 acres of wooded habitats, and over 72,000 acres of grasslands or farmlands. The only tourist accommodations are at Holdrege.

A. Federal Areas

1. Cottonwood Basin WPA (Map location 3). Area 79 acres wetland, 161 acres upland. Located 1 mile north and 2 miles east of Bertrand. Habitat includes 201 acres of wetland and 359 acres of upland. Recent Management: The eastern half of this WPA has been publicly owned for many years while the other half has been in private ownership. In 2000, the private portion of the wetland was sold to the U.S. Fish and Wildlife Service. Restoration plans include removal of the fence line/dike between the two halves, and reseeding the cultivated area back to native grasses. A pipeline is used to deliver water directly to the wetland. (Courtesy Eric Volden)

2. Linder WPA (Map location 4). Area 2 acres wetland, 79 acres upland.

3. Johnson Lagoon WPA (Map location 5). Area 252 acres wetland, 326 acres upland. Habitat includes 252 acres of wetlands and 326 acres of upland. The view is best from the east side, looking west. The mudflats on the west side are excellent for shorebirds. Whooping cranes and peregrine

falcons have been seen here in recent years during spring migration. Waterfowl and waterbirds are abundant. (Courtesy Eric Volden)

4. Funk Lagoon WPA (Map location 8). Area 1,163 acres wetland, 826 acres upland. Located 1 mile east and 3 miles north of Funk. This is the largest basin marsh at 1989 acres (1,163 wetland marsh acres and 826 upland acres), and perhaps the best. It is one of the few basins with permanent water and has some of the best marsh vegetation. During spring it hosts hundreds of thousands of geese (especially greater white-fronted), and some 20 species of ducks. Thousands of shorebirds use this site from March through May and again in early fall. In April and October, whooping cranes have used this area. From May through September you might see cattle egrets, black-crowned night herons, great blue herons and great egrets. White-faced ibis and cinnamon teal are regularly seen here. Pelicans, double-crested cormorants, and eared grebes are common in the deeper water areas. Birds that nest here include great-tailed grackles, yellow-headed blackbirds, eared and pied-billed grebes, least bitterns, Virginia rails, northern harriers and common yellowthroats. Playing birdsong tapes of sora and Virginia rails should elicit a response. The amount of surface water present each spring greatly affects waterfowl usage and natural runoff may be supplemented by groundwater pumping when needed. Funk lagoon includes large areas of open water, moist soil wetlands and restored native grasslands. Hiking trails along dikes offer excellent opportunities to view wildlife any time of the year. A 3-mile loop trail begins and ends at the main parking lot that has an information kiosk with maps and a nearby handicap-accessible observation blind that looks out over the marsh. The wetland is the collecting area for runoff from a large watershed. It can quickly go from nearly dry to flooded, after a heavy summer rain. Recent management has included prescribed burning, grazing, silt removal, disking, and reseeding of native grasslands. Dry conditions have allowed aggressive management of the areas choked with cattail and reed canary grass. (Courtesy Erick Volden). The main parking area has an information kiosk and a nearby observation blind looking out over the marsh. A variety of herons, egrets, and white-faced ibis visit the area in spring and fall.

5. Atlanta WPA (Map location 9). Area 453 acres wetland, 659 acres upland. Located 6 miles west and 3 miles south of Holdrege. Habitat includes 659 upland acres and 453 wetland acres, seasonally open to public hunting for pheasant, waterfowl and doves. Atlanta WPA contains a large wetland basin that requires a large runoff event to provide adequate water for migratory waterfowl. For this reason, the basin is dry in many low snowfall years. The property contains one well, which is unable to provide enough water (in relation to cost) to create suitable open water. In recent years, a couple of management practices have been attempted to improve the basin's use by waterfowl. In the mid-1990s, a low-level dike with a water control structure was built to separate the northern portion from the rest of the wetland. The wetland receives most of its runoff from the north. The diked area allows at least a portion of the basin to fill in low or marginal precipitation years. Intense grazing has also been done on the wetland, which has reduced the amount of vegetation, primarily reed canarygrass. The grazing allows even a couple of inches of water to become accessible to waterfowl. Nearly 70 acres of low diversity grassland (northeast corner) were inter-seeded with a high diversity seed mix. The upland has numerous volunteer trees scattered throughout the property. In 2002, trees were removed on approximately 85 acres. (Courtesy Eric Volden) A prairie dog town is present.

6. Jones Marsh WPA (Map location 10). Area 90 acres wetland, 76 acres upland. Located 3 miles west and 3 miles south of Holdrege. Habitat includes 76 acres of upland and 89 acres of wetland. Hunting is allowed. The wetland has been left idle for more than 10 years with little or no use by waterfowl. During that time, trees have established themselves along the wetland boundary. The amount of water held in the basin diminished and trees began growing in the center portion of the wetland. In December 2000, trees greater than three inches in diameter were removed. In the spring of 2002, a prescribed burn was done on the whole unit to burn up the tree piles and reduce the organic layer on the wetland bottom. Prescribed fire is expected to remove the smaller trees left standing. Waterfowl are expected to respond quickly to the change. In February 2003 an existing well was replaced and a new engine installed. Later in the

year, a pump house, covering the engine, was constructed. (Courtesy Eric Volden)

B. State Areas

1. West Sacramento WMA (Map location 11). Area 200 acres wetland, 188 acres upland. A prairie dog town of 4-5 acres is present.

2 Sacramento-Wilcox WMA (Map location 12). Area 1,050 acres wetland, 1263 acres upland. Located about 2.5 miles west of Wilcox. Sac offers a nice variety of habitat types, including freshwater marsh, prairie, creek and woodlands. Several controlled water impoundments insure some water is always available. A wide variety of birds can be seen here. Sacramento-Wilcox WMA was acquired by the Nebraska Game and Parks Commission in 1948. "Sac" serves as a waterfowl refuge and as a public hunting area. Approximately 500 of its 2,313 acres are designated as refuge and there is a recently constructed viewing blind, which overlooks a good waterfowl and shorebird area when water is present. Many ducks visit the area each fall, and good duck hunting is available from 19 established blinds as well as pheasant hunting. Intensive habitat development, including planting and managing trees, shrubs and grasses, has provided a wealth of cover diversity. Camping is available in a designated area. The headquarters is located on the east end of the property. Winter roosts of long-eared owl occur here. (Courtesy Eric Volden)

3. High Basin WMA (Map location 13). Located 2 miles north of Bertrand. This site includes 44 acres of wetland and 74 acres of pastureland. Good shorebird viewing, especially in April, May, July and August. The county roads nearest the Platte River are worth driving any time of the year. (Courtesy Eric Volden)

23. Dundy County

Dundy County is a high plains county with less than 500 acres of surface water, about 5,200 acres of wooded habitats, and 384,000 acres of grasslands or farmlands. There are no tourist accommodations in the county. Scattered areas of sandsage still exit in western areas.

A. Federal Areas: None

B. State Areas

1. Rock Creek Lake SRA. Area 165 acres. Located 10 miles west and 4 miles north of Benkelman, 1 mile south of fish hatchery. This may be the best birding area in southwestern Nebraska. The surrounding vegetation is one of the few protected areas of sandsage vegetation in Nebraska, with its associated birds such as Cassin's sparrow. This 54-acre reservoir is one of the few locations in the region where migrating water birds can settle, thus it attracts ducks, shorebirds and other water birds during both spring and fall. It also attracts many passerine migrants, especially in autumn, and migrating ospreys. Park entry permit required. http://outdoornebraska.ne.gov/parks/guides/park-search/showpark.asp?Area_No=152

24. Hitchcock County (Map 19)

Hitchcock County is a Republican Valley county with over 5,600 acres of surface water (mostly reservoir acreage), about 2,000 acres of wooded habitats, and over 200,000 acres of grasslands or farmlands. There are tourist accommodations at Trenton and Culbertson.

A. Federal Areas: None

B. State Areas

1. Swanson Reservoir & WMA (Map location 1), Area 1,157 acres upland, 4,973 acres reservoir. Primitive and modern camping facilities are present, and there are 13 hiking trails. Swanson Reservoir attracts many migrant water birds, some of which might overwinter. The wet meadows south of Stratton also attract many water birds during migration, including sandhill cranes and white-faced ibis (Rosche, 1994). About 3,000 acres are open to hunting and other public use; this is the largest of the area's reservoirs and has a large fish population, which should attract eagles and other fish-eating birds. A prairie dog town of about six acres is present.

25. Red Willow County

Red Willow County is a Republican Valley county with about 2,700 acres of surface water, 7,000 acres of wooded habitats, and nearly 180,000 acres of grasslands or farmlands. There are tourist accommodations at Indianola and McCook.

A. Federal Areas: None

B. State Areas

1. Red Willow SRA. Area includes 5,948 acres, including a 1,628-acre reservoir. Located 12 miles north and 2 miles west of McCook on US Highway 83. It consists mostly of high plains grasslands or farmlands. Complete camping facilities are present. The associated reservoir (Hugh Butler Lake) extends into Frontier County (see further information there). Park entry permit required. http://out-doornebraska.ne.gov/parks/guides/parksearch/showpark.asp?Area_No=149

2. Bartley Diversion WMA. One mile south and 1.5 miles east of Indianola. A small area of grasslands, rolling hills and scattered trees, around a campground.

C. Other Areas

1. Barnett Park. Located in McCook, on the north side of the Republican River. There is a small lake and a nature trail along the river.

26. Furnas County

Furnas County is a Republican Valley county with nearly 5,600 acres of surface water, 8,300 acres of wooded habitats, and over 175,000 acres of grasslands or farmlands. There are tourist accommodations at Arapahoe, Beaver City, Cambridge and Oxford.

A. Federal Areas: None

B. State Areas

1. Cambridge Diversion Dam. Located 2 miles east of Cambridge. Includes 21 acres of grassland bordering the Republican River and brushy bottomland.

27. Harlan County (Map 20)

Harlan County is a Republican Valley County with nearly 15,000 acres of surface water (mostly reservoir acreage), nearly 9,000 acres of wooded habitats, and almost 120,000 acres of grasslands or farmlands. There are tourist accommodations at Alma, Orleans, and Republican City.

A. Federal Areas

1. Harlan County Dam (Map location 1). Upland area 17, 278 acres, reservoir 13,338 acres. This largest reservoir in south-central Nebraska attracts bald eagles, geese (especially Canada geese), some sandhill cranes during spring and fall, and has a population of greater prairie-chickens (on the south side of the reservoir) as well. Near the south end of the dam is an eagle roost. Look for burrowing owls in the prairie dog colony between Republican City and the dam administration area that is located between the town and the dam. During spring and fall good numbers of double-crested cormorants and American white pelicans are present (some perhaps staying through summer), Gulls also accumulate in good numbers here, and migrating raptors are often seen.

B. State Areas

1. South Sacramento Wildlife Area (Map location 2). Area 77 wetland acres, 90 upland acres.

2. Southeast Sacramento Wildlife Area (Map location 3). Area 140 acres wetland, 45 acres upland. A prairie dog town of 4 acres is present.

Sharp-tailed Grouse

The East-Central Region: Sandhill Crane Country

The central Platte Valley and nearby Rainwater Basin provides some of the best spring birding opportunities in all of North America; for most of March about seven million waterfowl and about half a million sandhill cranes pour into the region, remaining until late March in the case of the waterfowl and about the second week of April in the case of the sandhill cranes. As the last sandhill cranes are leaving, whooping cranes begin to arrive, as do the earlier shorebirds, continuing the amazing spring spectacle until about the end of April.

Birding in the central Platte Valley during March is a chancy affair in terms of weather; late winter snowstorms may blanket the entire area in a foot of snow, which when melting leaves country roads slippery at best, and thus driving requires a good deal of care. This is especially true in the Rainwater Basin, an area of clay soils that prevent water from percolating down, and thus is rich in temporary wetlands (locally called "lagoons") just at the peak of spring waterfowl populations. This is only true during years when winter snowfalls or spring rains allow the basins to fill; in drier years only the deepest lagoons or those that are kept wet by pumping (Harvard, Massie's, Smith, etc.) can accommodate the hordes of ducks and geese passing through. During such years the stresses caused by bad weather and overcrowding can set off outbreaks of fowl cholera, and kill tens of thousands of birds in only a short time. Some of these birds are consumed by wintering bald eagles, hundreds of which occur along ice-free areas of the Platte from late fall until early spring. A good viewing area for these birds is at the J-2 Hydro Plant near Lexington. This area is open to the public on Saturdays and Sundays from 8 a.m. to 2 p.m. with weekday reservations possible for groups (call 308/995-8601 for information).

The best way to watch cranes during the day is observing them field-feeding from a parked car, with observers remaining quiet and inside the car.

Opening a door and leaving the car will guarantee a rapid departure of the birds. Gravel roads on the south side of the Platte River are usually better than those on the north side of Interstate 80. The most rewarding way to watch cranes is from riverside blinds near roosting locations (see accompanying maps). Such blinds are maintained by the Whooping Crane Trust on Mormon Island (reservations required; cost $20 per person, or the Audubon Society's Lillian Rowe Sanctuary near Gibbon (cost $20 per person, for reservations phone 303/468-5282), and possibly the Fort Kearny State Historical Park near Kearney (reservations required, 308/865-5305). If it is not possible to arrange a blind viewing, several bridges such as the hike-bike trail bridge near Fort Kearney or the bridge over the middle Platte channel two miles south of Alda (see maps 26 & 27) provide a less thrilling but still exciting view, both at sunset and sunrise. Information on crane viewing and accommodations can be obtained from the Kearney Visitors Center (308/652-9435 in state; 800/227-8340 out-of-state), the Grand Island Visitors Bureau (800/658-3178 or 308/382-4400), or the Adams County Visitors Bureau in Hastings (800/967-2189 or 402/461-2370). The Hastings Museum (14th St. & Burlington Highway, 402/461-4629) and the Stuhr Museum at the southern edge of Grand Island along US Route 34 (308/385-5316) both provide tourist information and sell informative books or pamphlets on local tourist attractions.

An excellent source of both general and specific information on birding in the Platte Valley is available in Gary Lingle's book *Birding Crane River: Nebraska's Platte*, which is locally available. It also includes complete county maps and detailed bird-finding advice for seven Platte Valley counties. A more general nature guide, *A Guide to the Natural History of the Central Platte Valley of Eastern Nebraska*, is available from P. Johnsgard. Other regional natural histories by P. A. Johnsgard are the University of Nebraska Press book *The Platte: Channels in Time,* the Smithsonian Institution Press book *Crane Music: The Natural History of American Cranes,* and *Those of the Gray Wind: The Sandhill Cranes,* University of Nebraska Press. The Nebraska Game & Parks Commission (308/865-5308 in Kearney, 402/471-0641 in Lincoln, or PO Box 30370, Lincoln, NE 68503) can provide free informative materials, including an excellent 8-page "Spring migration Guide" that centers on Platte Valley birding. Some maps showing wildlife viewing areas are reproduced here by permission. The Game and Parks Commission has also (1997) published a 96-page booklet by Joseph Krue, titled *NEBRASKAland Magazine Wildlife Viewing Guide,* which includes descriptions of 68 sites in the state. It may sometimes be obtained from used bookstores or possibly from its publisher, Falcon Press. A collection of county road maps is also available from the Nebraska Department of Roads in Lincoln (402/471-4567); bound sets of maps covering the entire state can be found in the *Nebraska Sportsman's Atlas* and the *Nebraska Atlas and Gazetteer.* Both describe and list local tourist attractions, offer camping information, and provide other similar information; the latter book being of contour maps and the former emphasizing hunting and fishing sites.

The Rainwater Basin area is just as attractive as the Platte Valley during early spring, when snow meltwaters accumulate in the clay-rich lowlands and an estimated 7-9 million ducks and 2-3 million geese pass through. These flocks include 90 percent of the mid-continental greater white-fronted goose population, 50 percent of the mid-continental mallard population, and 30 percent of the entire continent's northern pintail population. Increasing numbers of snow geese also use the more easterly parts of the area each spring, the numbers recently exceeding a million birds.

Some of the shallower wetlands are also of great importance to migrant shorebirds. A recent (2004) monograph by Joel Jorgensen analyzed shorebird migration patterns in the eastern Rainwater Basin. He reported that, in decreasing order, the most numerous shorebirds (out of 38 total species) seen there over a several-year period were white-rumped sandpiper, Wilson's phalarope, semipalmated sandpiper, long-billed dowitcher, lesser yellowlegs, least sandpiper and Baird's sandpiper during spring. Pectoral sand-

piper, long-billed dowitcher, lesser yellowlegs, least sandpiper and stilt sandpiper were progressively less numerous during fall. The region may be of hemispheric migratory importance to the very localized buff-breasted sandpiper, which stages in various mixed-grass sites around the eastern Rainwater Basin.

A suggested loop tour (ca. 135 miles) is shown by arrows on Maps 28 and 29, starting from I-80 exits at either York or Aurora, and returning to the opposite exit. This route over generally improved roads includes the most important wetlands of the eastern Rainwater Basin. The Rainwater Basin Wetland Management District comprises about 84 wetlands occupying 28,600 acres (including 21,742 acres federally owned, and about 6,900 acres state owned). Sites such as Harvard, Massie and Smith's Lagoons, and Mallard Haven, are of special value to these birds and are prime birding locations in the eastern basin, while Funk Lagoon is of special attraction in the western basin. Sites of special shorebird significance include Harvard, Mallard Haven, Massie, and Sinninger. A bird checklist, with nearly 200 species including over 100 breeding species, is available from the Kearney (Rainwater Basin) office of the USFWS, 2610 Ave. Q, Kearney 68848 (308/236-5015). The Kearney office of the Game & Parks Commission is 1617 First Ave., Kearney (308/865-5310).

1. Boyd County

Boyd County is a Niobrara Valley county with about 2,500 acres of surface water, 1,600 acres of wooded habitats, and 200,000 acres of grasslands or farmlands. The only tourist accommodations are at Spencer.

A. Federal Areas: None

B. State Areas

1. Parshall Bridge WMA. Area 230 acres. Located five miles south of Butte. Riparian wooded habitats along the Niobrara River.

2. Hull Lake WMA. Area 36 acres. Located three miles south and one mile west of Butte. Hilly uplands, with oaks, conifers and grasslands around three-acre lake.

2. Holt County

Holt County is a mostly Sandhills county with over 12,000 acres of surface water, almost 69,000 acres of wooded habitats, and 1.2 million acres of grasslands or farmlands. The only tourist accommodations are at Atkinson and O'Neill.

A. Federal Areas: None

B. State Areas
1. Atkinson Lake SRA. Area 54 acres. Located at northwest edge of Atkinson. Includes a 14-acre reservoir of the Elkhorn River. State park entry permit required.
http://outdoornebraska.ne.gov/parks/guides/park-search/showpark.asp?Area_No=9

2. Goose Lake WMA. Area 349 acres. Located four miles west of highway 281 and two miles north of Wheeler County boundary (23 miles south and four miles east of O'Neill). Mostly lake; also grassy and wooded uplands.

3. Redbird WMA. Area 433 acres. Located one mile south of highway 281 bridge over Niobrara. Mostly bur oak and cedar; bisected by Louse Creek, with steep wooded slopes and rolling grasslands.

4. Spencer Dam WMA. Located 23 miles north of O'Neill on U.S. 83. Includes 9 acres of Niobrara River Valley, providing river access.

3. Knox County (Map 21)

Knox County is bounded by the Niobrara and Missouri rivers, and thus has over 41,000 acres of surface water, as well as 38,000 acres of wooded habitats, and almost 320,000 acres of grasslands or farmlands. There are tourist accommodations at Bloomfield, Creighton, Crofton, Niobrara and Wausa.

A. Federal Areas: None

B. State Areas
1. Gavin's Point Dam, Lewis & Clark Lake SRA (Map location 1). Area 32,000 acres reservoir & 1,227 acres SRA. See Cedar County account.
www.nwo.usace.army.mil/html/Lake_Proj/gavin-spoint/

2. Bazille Creek WMA (Map location 2). Area 4,500 acres. Bordered for 9 miles by the Missouri River and Lewis and Clark Lake; mixed woods, grasslands and marshy areas. This area is extensively marshy, as it includes the area where the Missouri River is impounded to form the upper end of Lewis and Clark Lake, so many wetland birds are present.

3. Niobrara State Park (Map location 3). Area 1,632 acres. This state park is located at the confluence of the Niobrara River and the backwaters of the Missouri River. It is mostly grassland, but also has riparian wooded habitats. There are more than 12 miles of hiking trails, and a new two-mile hike-bike trail extends along the park's northern boundary. Wooded habitat birds include whip-poor-wills, and both bald eagles and ospreys are seasonally present. There is also an interpretive center, and both modern cabins and primitive camping facilities are present. Classified as a Nebraska Important Bird Area. State park entry permit required; phone 402/857-3373 for further information.
http://outdoornebraska.ne.gov/parks/guides/park-search/showpark.asp?Area_No=126

4. Bohemia Prairie WMA (Map location 4). Area 680 acres. Mainly grasslands, with some woods and two ponds.

5. Greenvalle WMA. Located ten miles west and three miles south of Verdigre. Area 200 acres, mostly wooded, and bisected by Middle Verdigre Creek.

4. Antelope County (Map 22)

Antelope County is an Elkhorn Valley county with less than 900 acres of surface water, over 21,000 acres of wooded habitats, and nearly 200,000 acres of grasslands or farmlands. There are tourist accommodations at Elgin, Neligh and Orchard.

A. Federal Areas: None

B. State Areas
1. Ashfall Fossil Beds State Historical Park. (Map location 1). Area 360 acres. This extremely important paleontological site preserves the fossils of horses, rhinos, camels and other animals

(including crowned cranes) killed and interred under a thick layer of volcanic dust that settled here about ten million years ago. The center is open from Memorial Day through Labor Day, 9-5 on weekdays and 11-5 on Sundays. There are shorter hours during May and September. The area is mostly rugged range country, with grassland species most common. However, rock wrens often can be seen near the excavation site. Admission fee, and state park permit required. For information write PO Box 66, Royal, NE 68773 (402/893-2000).
http://outdoornebraska.ne.gov/parks/guides/park-search/showpark.asp?Area_No=279

2. Grove Lake WMA (Map location 2). Area 1,746 acres, 35-acre reservoir. Mainly mixed hardwoods and grasslands along Verdigre Creek. This is rolling grassland, with scattered trees along East Verdigre Creek. There is also a small reservoir and a trout-rearing facility. Most of the birds are grassland forms, but ospreys and belted kingfishers are also possible. Phone 402/370-32374 for information.

3. Hackberry Creek Public Use Area (Map location 3). Area 180 acres. Includes a mile of Elkhorn River frontage, several marshy oxbows, mixed woods and grassland.

4. Redwing WMA (Map location 4). Area 320 acres. Includes 1.5 miles of Elkhorn River frontage. Mostly riparian wooded habitats, some grassland and marshes.

5. Pierce County (Map 23)

Pierce County is a county with less than 4,000 acres of surface water, about 5,000 acres of wooded habitats, and nearly 125,000 acres of grasslands or farmlands. There are tourist accommodations at Osmond and Plainview.

A. Federal Areas: None

B. State Areas
1. Willow Creek SRA. (Map location 1). Area 1,600 acres, with a 700-acre reservoir. A fishing and camping area. State park entry permit required.

6. Garfield County

Garfield County is a Sandhills County with about 6,000 acres of surface water (nearly all reservoir acreage), 5,500 acres of wooded habitats, and over 320,000 acres of grasslands or farmlands. There are tourist accommodations at Burwell.

A. Federal Areas: None

B. State Areas
1. Calamus Reservoir SRA (10,312 acres, including 5,123 acres reservoir). Includes several modern camping facilities, boating and related facilities, plus a fish hatchery. Classified as a Nebraska Important Bird Area. See state highway map for location; see also Loup County. State park permit required.
http://outdoornebraska.ne.gov/parks/guides/parksearch/showpark.asp?Area_No=275

7. Wheeler County

Wheeler County is an eastern Sandhills county with 1,300 acres of surface water, almost 24,000 acres of wooded habitats, and nearly 330,000 acres of grasslands or farmlands. There are tourist accommodations at Erickson.

A. Federal Areas: None

B. State Areas
1. Pibel Lake SRA. Area 42 acres, plus a 45-acre lake. Located seven miles east and two miles south of Erickson; see state highway map. This is a beautiful Sandhills lake with a variety of water birds present during spring and summer. State park entry permit required.
http://outdoornebraska.ne.gov/parks/guides/park-search/showpark.asp?Area_No=138

8. Valley County

Valley County is a Loup Valley county with nearly 3,000 acres of surface water, 2,300 acres of wooded habitats, and 200,000 acres of grasslands or farmlands. There are tourist accommodations at Ord.

A. Federal Areas: None

B. State Areas

1. Fort Hartsuff State Historical Park. Located about three miles north and one mile west of Elyria. See state highway map for exact location. Mainly of interest for historical reasons, but passerines should be present seasonally. Open April through October (308/346-4715). State park permit required.
http://outdoornebraska.ne.gov/parks/guides/park-search/showpark.asp?Area_No=74

2. Davis Creek WMA and SRA. Located about three miles south of North Loup. Mostly grassy areas surrounding a 1,145-acre reservoir. The WMA includes about 2,000 acres exclusive of the recreation area.

3. Scotia Canal WMA. Located 4.5 miles north of North Loup. Near North Loup River, and mostly covered by grassy uplands and mixed wooded habitats.

9. Greeley County

Greeley County is an eastern Sandhills and loess hills county with nearly 2,500 acres of surface water (mostly reservoir acreage), almost 3,000 acres of wooded habitats, and over 200,000 acres of grasslands or farmlands. There are tourist accommodations near Spalding.

A. Federal Areas: None

B. State Areas
1. Davis Creek SRA. Area 2,000 acres, plus a 1,145-acre reservoir. See Valley County description.

10. Boone County

Boone County is a mostly loess hills county with about 1,100 acres of surface water, 2,600 acres of wooded habitats, and over 160,000 acres of grasslands or farmlands. There are tourist accommodations at Albion.

A. Federal Areas: None

B. State Areas
1. Beaver Bend WMA. Area 27 acres. Located one mile northwest of St. Edward. See state highway map. An area located along Beaver Creek, with riparian wooded habitats.

11. Madison County (Map 23)

Madison County is a loess hills and Elkhorn Valley county with less than 800 acres of surface water, about 10,000 acres of wooded habitats, and over 80,000 acres of grasslands or farmlands. There are tourist accommodations at Newman Grove, Norfolk and Tilden.

A. Federal Areas: None.

B. State Areas
1. Yellowbanks WMA (Map location 2). Area 680 acres. Includes 1.5 mile frontage of the Elkhorn River. Includes steep riverine bluffs supporting mature hardwood forest, some grassy uplands.

2. Oak Valley WMA(Map location 3). Area 640 acres. Includes a hardwood bottomland forest, bisected by Battle Creek; otherwise grassy uplands.

12. Platte County (Map 25)

Platte County is a mostly loess hills county with almost 3,000 acres of surface water, 7,400 acres of wooded habitats, and nearly 90,000 acres of grasslands or farmlands. There are tourist accommodations at Columbus and Humphrey.

A. Federal Areas: None

B. State Areas
1. George Syas WMA (Map location 4). Area 917 acres. Includes 1.5 miles of Loup River frontage. About half wooded, the rest grasses, crops and planted shrubs.

2. Wilkinson WMA. 957 acres. Upland grassland and managed wetlands. Located 5 miles west and 1 mile north of Columbus, just off Highway 81. Large numbers of waterfowl and shorebirds in spring. Not shown on map.

3. Looking Glass Creek WMA. Located 11 miles south of Monroe. About half wooded, the rest grassland, with two small lakes.

C. Other Areas

1. Lake Babcock Waterfowl Refuge and Lake Babcock (Map location 6). Area 600 acres. A reservoir on Shell Canal, just outside of Columbus.

2 .Lake North (Map location 7). Area 200 acres. A city lake developed for fishing and swimming; probably of limited birding potential.

3. Pawnee Park. City park on banks of Loup River. Good riparian zone for spring warblers.

13. Sherman County (Map 24)

Sherman County is a Loup Valley County with over 4,000 acres of surface water (mostly reservoir acreage), 3,000 acres of wooded habitats, and over 200,000 acres of grasslands or farmlands. There are tourist accommodations at Loup City.

A. Federal Areas: None

B. State Areas
1. Sherman Reservoir SRA/WMA (Map location 1). Area 3,382 acres, plus 2,845-acre reservoir. Mostly rolling prairie grasslands, with woody growth along creeks. Includes 10 hiking trails. Two prairie dog towns totaling about 12 acres are present. State park entry permit required.

2. Bowman Lake SRA (Map location 2). Area 23 acres. A small SRA just outside Loup City. State park entry permit required.

14. Howard County

Howard County is a Loup Valley county with nearly 3,000 acres of surface water, over 5,000 acres of wooded habitats, and about 190,000 acres of grasslands or farmlands. There are tourist accommodations at Dannebrog and St. Paul.

A. Federal Areas: None

B. State Areas
1. Harold W. Andersen WMA. Area 272 acres. Located four miles south and two miles west of St. Paul. Consists of about 12 miles of Loup River frontage, with bottomland timber and a marshy oxbow.

2. Loup Junction WMA. Area 328 acres. Located three miles north and two miles east of St. Paul. Bordered on the north by the North Loup and on the south by the Middle Loup; mostly riparian wooded habitats, with marshes and grassy areas.

15. Nance County (Map 25)

Nance County is a Loup and Platte Valley county with almost 3,000 acres of surface water, over 5,000 acres of wooded habitats, and over 190,000 acres of grasslands or farmlands. There are tourist accommodations at Genoa and Fullerton.

A. Federal Areas: None

B. State Areas
1. Loup Lands WMA (Map locations 1, 2). Area 485 acres.

2. Prairie Wolf WMA (Map location 3). Area 972 acres. Mainly bottomlands along the Loup River, with restored grasslands and marshes, and some timber.

3. Sunny Hollow WMA (Map location 5). Area 160 acres. Mostly grassy uplands, with two marshes and a "dugout" wetland.

4. Council Creek WMA. 160 acres of alfalfa, restored grassland and riparian wood. Located 6.5 miles west and 1 mile south of Genoa. Not shown on map.

16. Polk County

Polk County is a Platte Valley county with less than 500 acres of surface water, nearly 5,000 acres of wooded habitats, and 54,000 acres of grasslands or farmlands. There are tourist accommodations at Osceola.

A. Federal Areas: None.

B. State Areas: None

17. Buffalo County (Map 26)

Buffalo County is a Platte and Loup Valley county with 4,400 acres of surface water, 9,600 acres of wooded habitats, and nearly 225,000 acres of grasslands or farmlands. There are tourist accommodations at Elm Creek, Gibbon and Kearney.

A. Federal Areas: None

B. State Areas

1. Ravenna Lake SRA (Map location 1). Area 53 acres. Situated along the South Loup River, with a small reservoir. State park entry permit required.

2. Blue Hole WMA (Map location 2). Area 530 acres, plus 30-acre pond and two miles of river frontage. Mostly riparian wooded habitats.

3. Sandy Channel SRA (Map location 3). Area 133 acres; 11 small lakes and ponds, totaling 47 acres. State park entry permit required.

4. Union Pacific SRA (Map location 4). Area 26 acres, plus a 15-acre pond. State park permit required.

5. East Odessa WMA (Map location 5). Area 71 acres, and a seven-acre pond. State park permit required.

6. Cottonmill Lake Public Use Area. (Map location 6). A hike-bike trail extends six miles from this area to the outskirts of Kearney.

7. Bassway Strip WMA (Map location 10). 636 acres, 4 ponds and 7 miles of river frontage. Includes 90 acres of lakes and sandpits; mostly wooded. In spite of the river frontage, this area is not used by sandhill cranes to any great extent.

8. War Axe SRA (Map location 8). Area nine acres plus a 12-acre pond. State park permit required. http://outdoornebraska.ne.gov/parks/guides/park-search/showpark.asp?Area_No=183

9. Windmill SRA (Map location 9). Area 168 acres, 5 ponds. State park entry permit required. http://outdoornebraska.ne.gov/parks/guides/park-search/showpark.asp?Area_No=196

C. Other Areas

1. Lillian Annette Rowe Sanctuary & Iain Nicolson Audubon Center (Map location 7). Area ca. 2,000 acres, counting conservation easements. This area, the largest Audubon refuge in the state, protects prime sandhill and whooping crane habitats near Kearney, and includes nearly six miles of river frontage, plus about 260 acres of native prairie. Several riverside blinds are located on the property, and spring sunrise (5 a.m.) or sunset (5 p.m.) excursions to the blinds can be arranged between early March and mid-April ($20.00 per person, reservations are needed). There is also a self-guided hiking/birding trail. The sanctuary headquarters provides information and sells books and other bird-related materials. Summer breeding birds include dickcissel, upland sandpiper and bobolink, as well as riparian wooded habitats and species such as rose-breasted grosbeak and willow flycatcher. Least terns and piping plovers often nest on barren sandbars that are also used by roosting cranes (Lingle, 1994). The bird list shown for the Platte Valley in the supplement should suffice for this site. Classified as a Nebraska Important Bird Area. The Iain Nicolson Center is of modern hay-bale construction; its north side is lined with windows for easy bird-watching. The sanctuary's address is 44450 Elm Island Rd., Gibbon, NE, 68840 (308/468-5282), email address: rowe@nctic.net. Office hours are 9-5 from Monday through Friday, Sunday 1-5 p.m., open 7 days a week during crane season). No admission charge, but $2.00 donation requested. www.rowesanctuary.org

2. Prairie-chickens often have a lek visible from the road near Kearney. Drive west 6.5 miles from town center, turn right (north) on Evergreen Road, go to "56" road and turn left (east). At 3.2 miles you will be at the top of a hill with two windmills visible. The lek is on the north side of the road, about 300 yards away, so a spotting scope is needed.

3. Richard Plautz Viewing Site (Map location 11). This bridge, 1.5 miles south of I-80 exit 285 in Buffalo County, provides a parking area and viewing platform (Central Platte Natural Resource District, Richard Plautz Viewing Site) for watching crane roosting flights. The Central Platte Natural Resources District (CPNRD) has led a task force in providing a series of crane viewing decks for use by visitors. The decks provide a safe area to view cranes and other wildlife on the Platte River. Best times for viewing are sunset and sunrise. Cranes, herons, egrets, pelicans, waterfowl, song sparrows

and a host of other birds can be seen comfortably from this wooden observation deck. Be careful to avoid standing on the bridge itself, where traffic can be quite fast. (Courtesy Eric Volden)

18. Hall County (Map 27)

Hall County is a Platte Valley county with nearly 2,000 acres of surface water, 3,900 acres of wooded habitats, and almost 120,000 acres of grasslands or farmlands. There are tourist accommodations at Alda, Grand Island and Wood River.

A. Federal Areas

1. Hannon WPA. 659 acres. Located 1 mile east and 2 miles north of the I-80 Shelton Exit 292. Habitat includes wet meadows and surrounding grassy uplands with 105 acres of water. Common summer residents include marsh and sedge wrens, upland sandpipers, bobolinks, dickcissels and a variety of native sparrows. This site has had good use by waterfowl when water is present and excellent use by sandhill cranes after a prescribed burn. There are several small ponds and a slough that runs through it on wet years which is attractive to shorebirds such as snipe. Some years this area has seen heavy use by migrating sandhill cranes. (Courtesy Eric Volden)

B. State Areas

1. Cornhusker WMA (Map location 2). Area 840 acres. All upland habitats with various planted cover types. The birds include such brush-loving winter species as Harris' sparrows and American tree sparrows.

2. Mormon Island SRA (Map location 10). Area 152 acres, 61-acre lake. Located 0.25 miles north of I-80 at the Grand Island Exit 312. Habitat includes 3 lakes and their surrounding riparian woodlands. Camping, restrooms, swimming, shelters and an office on site. This area is a popular fishing, camping, and swimming spot just off I-80 which occasionally attracts large concentrations of waterfowl in the spring despite the potential for heavy disturbance. Used as a winter stopover by Mormon emigrants heading westward, Mormon Island State Recreation Area is part of Nebraska's unique "Chain of Lakes." Development of I-80 in the early 1960s created a series of "borrow pit" lakes when sand and gravel were removed for road construc-

tion. The first of these areas developed was Mormon Island SRA. A good variety of waterfowl and shorebirds come to the area before heading farther north. Ask park personnel for more information about the best times and locations to view these impressive wildlife displays. Because of its depth, the main lake sometimes hosts loons, pelicans, mergansers and a variety of grebes. The slough running through the SRA is a good place to search for snipe. Cedar waxwings, woodpeckers and brown creepers are common, as are owls. (Courtesy Eric Volden) This area is a popular fishing spot just off I-80, and rarely attracts many waterfowl because of the high disturbance level.

http://outdoornebraska.ne.gov/parks/guides/park-search/showpark.asp?Area_No=123

3. Martin's Reach WMA. Located one mile south and three miles west of Wood River Exchange. Includes 89 acres, with about 0.7 miles of river frontage of the middle channel of the Platte River. A slough running through the center provides nesting habitat for shorebirds and ducks. As many as 88 species have been seen here in a single day.

4. Loch Linda WMA. 38 acres. Located 3 miles east of the Alda I-80 exit 300, then a mile south over the interstate and 2 miles east. This is a 29-acre wet cattail marsh surrounded by 9 acres of pastureland and mature riparian forest adjoining the Platte River. Ducks, wild turkey, yellow-headed blackbirds, common yellowthroats and herons are common in the marsh and shorebirds should be visible along the Platte. (Courtesy Eric Volden).

C. Other Areas

1. Taylor Ranch road (Map location 1). Located 4 miles west and 3 miles north of Grand Island. Taylor Ranch is a privately owned 7,000-acre ranch with extensive Sandhills prairie and numerous small wetlands that are attractive to migrating ducks and shorebirds during wet years. County roads along the perimeter of this ranch provide an opportunity to watch displaying greater prairie-chickens from a parked car. Active prairie-chicken leks can be located by driving this area around sunrise and stopping every few hundred yards or so to listen for their "booming" from mid-March into May. A few sharp-tailed grouse are also present and a good variety of raptors can be found too. Around 90 species have been observed here. Blue

grosbeaks have nested in the plum thickets and burrowing owls nested here in 2004. (Courtesy Eric Volden). The arrowhead on the map shows one such location where a lek usually is located. Farther north in the Sandhills sharp-tailed grouse outnumber the greater prairie-chickens.

2. The Crane Trust. (Map location 3). This preserve of about 2,500 acres was the first Platte Valley crane sanctuary to be established, and along with the Rowe Sanctuary farther west is the most important. More than 70,000 cranes have been seen on its pristine wet meadows, and up to 80,000 birds roost along its river shorelines. Nearly 220 bird species have been reported here; the Platte Valley bird list in the supplement is largely based on these records. This facility is generally not open to the public. For information on The Crane Trust call 308/384-4633.
www.nebraskabirdingtrails.com/site.asp?site=363

3. Shoemaker Island road. (Map location 4). Located 2 miles south of the I-80 Alda Exit 305, running west to 1 mile south of the Wood River I-80 Exit 300. This gravel road traverses the length of Shoemaker Island, where many wet meadows attract foraging flocks of cranes. There are also large stands of riparian forest where rose-breasted and black-headed grosbeaks can be observed along with eastern wood-pewees, wild turkey and red-headed woodpeckers. The entire area is privately owned, so birding away from the road requires landowner permission. Cattle egrets, black-billed magpies and eastern bluebirds are common here. Road ditches often contain some water and wood duck, sora rails and American bitterns sometimes make use of them. Greater prairie-chickens have been infrequently observed on this island. American woodcock perform their courtship skydance along the wooded river drainage in April and May. (courtesy Eric Volden) This stretch of the Platte River is used by migrating sandhill cranes, waterfowl, eagles, shorebirds, and by nesting piping plovers and least terns. At least 205 species of birds have been recorded in this area. The adjacent native prairie provides nesting sites for prairie species such as dickcissels, upland sandpipers, bobolinks, grasshopper sparrows and Bell's vireos. Riparian areas provide habitat for a variety of passerines including black-headed grosbeaks, orchard orioles, willow flycatchers, and black-billed

magpie, while supporting a thriving wild turkey population.

4. Nebraska Nature and Visitor Center (Map location 5). This newly reorganized nature center was previously known as Crane Meadows and Nebraska Bird Observatory. It is now jointly operated by the Grand Island Tourist Bureau and Hastings College. Its website is http://nebraskanature.org, and may be reached by e-mail at: brad@nebraska-nature.org. Like the previous nature center, guided blind tours may be arranged during the crane season (early March to early April), and there are staff available to help provide birding information. Hiking trails are open year round, with some restrictions during crane season. The Center's address is P.O. Box 90, Alda, NE.

5. Alda Road Bridge (Map location 6). This bridge over the middle Platte (2 miles south of I-80 exit 305) channel provides a place (CPNRD Alda Site) where people can watch the sunrise and sunset roosting flights of cranes. It is very near a sandpit lake that may attract up to 40,000 geese, but this lake is on private property and can only be viewed from the highway. Located in Hall County, 2 miles south of I-80 exit 305 on the Alda Road. The handicap-accessible viewing platform provides an excellent view of cranes and waterfowl roosting on both sides of the Platte's Aida Bridge at sunrise and sunset. This site has a small lake and a paved hiking trail. It is adjacent to a private sandpit lake that may attract over 100,000 geese. The Central Platte Natural Resources District (CPNRD) led a task force that built several free crane viewing decks that provide a safe area to view cranes and other wildlife on the Platte River. Parking is available and best times for viewing are sunset and sunrise. Besides cranes, herons, egrets, pelicans, waterfowl, song sparrows and a host of other birds can be seen comfortably from this wooden deck. (Courtesy Eric Volden)

6. Platte River Road. (Map location 7). This paved road going west from Doniphan is a good route for observing field-feeding cranes during the daytime. It continues west to the Kearney area, but the density of crane use varies with location and disturbance. Generally the cranes are best seen from the road nearest the south shore of the Platte River, especially in early morning and late

afternoon, among cornfields or the occasional wet meadows that still exist.

7. Amick Acres road. (Map location 8). This small subdivision has several sandpit lakes that attract large flocks of Canada and cackling geese in early March. Do not stray from the road, as the area is entirely private property.

8. Nine-mile Bridge. (Map location 9). This narrow bridge north of Doniphan provides views of a small crane flock on the downstream side. However, no parking is allowed near the bridge, and so some walking is necessary.

9. Hall County Park & Stuhr Museum (Map location 12). Located in Hall County, South of Grand Island near the Stuhr Museum, 1 mile south of Hwy 34 on U.S. HWY 281 and 0.5 miles east on Shimmer Drive. This county park allows free entry and offers wooded trails for birding along with camping facilities. Warblers, thrushes, woodpeckers and kinglets are seen here. Occasionally a Carolina wren, American redstart, or long-eared owl can be found here too. This heavily wooded 38-acre park is the remnant of dried-up Shimmer Lake. The Wood River forms its northern boundary. The county park allows free entry, and offers wooded trails for birding. The museum has an admission charge, but there is free access to the museum shop. For information call 308/385-5316. (Courtesy Eric Volden)

10. Eagle Scout Park, 90 acres. Located 3 miles north of Grand Island on Highway 281. Eagle Scout Park has an 80-acre lake surrounded by a 1.2 mile paved hiking trail. Trees, shrubs, and mowed grassy areas border the lake. Playground and restroom facilities are available along with parking lots on the south, east and north sides. A well used haven for waterfowl, shorebirds and waders including egrets, avocets, mergansers and pelicans. Look for sparrows and warblers in the surrounding trees and brush including song sparrows, Nelson's sharp-tailed sparrow (migration) and yellow warblers. (Courtesy Eric Volden)

11. Mormon Island Crane Meadows. 2,500 acres. Located 1 mile south of I-80 exit 312 on U.S. HWY 281 then west on Elm Island Road. This is the largest remaining wet meadow left along the Platte

River and is owned and managed by the Whooping Crane Trust. This 2,500-acre preserve contains an extensive area of sedge meadows along with native tall grass prairie surrounded by channels of the Platte River. This was the first Platte Valley crane sanctuary to be established. More than 70,000 cranes have been seen foraging together on its pristine wet meadows and as many as 80,000 sandhill cranes roost along its river shorelines during peak usage. Two hundred and twenty-three bird species have been reported here. Good numbers of upland sandpipers, bobolinks, sedge wrens, dickcissels and grasshopper sparrows nest here from May till August. Prairie falcons, short-eared owls and northern harriers are seasonally common here. Access is by permission only. For information call 308/384-4633. (Courtesy Eric Volden).

19. Merrick County (Map 25)

Merrick County is a Platte Valley county with about 600 acres of surface water, over 13,000 acres of wooded habitats, and more than 113,000 acres of grasslands or farmlands. There are tourist accommodations at Central City.

A. Federal Areas: None

B. State Areas:
1. Dr. Bruce Cowgill WMA. 150 acres. Platte River frontage, riparian timber, planted grasslands and wet meadows. Located 1.5 miles east of silver Creek, just south of Highway 30. Not shown on map.

C. Other Areas
1. Bader Memorial Park Natural Area. Located three miles south of Chapman. Area 80 acres. A stretch of Platte River wooded habitats and adjacent native prairie, with trails through all of the local habitat types. American woodcocks occur here, and sandhill cranes sometimes visit during spring. Ducks, geese, marsh birds and shorebirds are abundant during migration. Entry fee.

20. Hamilton County (Map 28)

Hamilton County is a Platte Valley and Rainwater Basin county with less than 1,000 acres of permanent surface water, 2,400 acres of wooded habitats, and almost 50,000 acres of grasslands or farmlands. There are tourist accommodations at Aurora.

A. Federal Areas:

1. Springer WPA. (Map location 3). Area 266 wetland acres, 134 acres upland. Located 6 miles west and 1 mile south of Aurora. Habitat includes 397 acres of wetland and 243 acres of upland. The Fish and Wildlife Service purchased the first tract in 1991, with the last purchase occurring in 1995. This basin was completely farmed prior to acquisition and represents the first major restoration project in the rainwater basin. Beginning in 1997, the south 160 acres were reseeded with a high diversity mix of grasses, forbs, sedges and rushes. Livestock grazing is used to control the invasion of reed canary grass. The large, flat wetland provides excellent habitat for waterfowl and other water birds. A well and pump are located on the property which is routinely pumped with water in the spring and fall. (Courtesy Eric Volden)

2. Troesler Basin WPA (Map location 4 right). Area 123 wetland acres, 37 upland acres.

3. Nelson WPA (Map location 5). Area 143 wetland acres, 17 acres upland.

B. State Areas

1. Pintail WMA (Map location 4 left). Area 190 wetland acres, 94 upland acres. Located 2.5 miles south and 2 miles east of the Aurora I-80 exit 332. Habitat includes 268 acres of marsh, 185 acres of cropland and 25 acres of pasture land. A large basin southeast of Aurora includes a shallow seasonal pond and mixed upland and lowland habitats. In wet springs this shallow marsh may attract up to 100,000 geese, primarily white-fronted, and is a favorite stopover for pintails and white-fronted geese. In the mornings, a new parking lot on the east side provides the best viewing access to the marsh. During afternoon the west side is better and the road is closer to the marsh. Some people use the old duck hunting dugouts on the southwest corner as viewing blinds. Look for pheasants, northern harriers and migrating peregrines in the uplands. Shorebirds, waders, American white pelicans, black terns, and a variety of waterfowl rest here during spring migration. (Courtesy Eric Volden)

2. Gadwall WMA (Map location 1). Area 90 acres, with 70 acres of wetlands (two "dugout" wetlands and narrow slough).

3. Deep Well WMA (Map location 2). Area 78 acres, with 35 acres of semi-permanent wetlands and 25 acres of permanent wetlands. Located 3 miles south of Phillips. Habitat includes 35 acres of semi-permanent wetlands and 25 acres of permanent wetlands. 70 acres of marsh, 43 acres of pasture and 125 acres of cropland. Known locally as the Phillips basin. Mudflats and emergent vegetation harbor semipalmated plover, marbled godwit, willet, black tern, and a host of other waterbirds in May. Waterfowl concentrations peak in mid-March. Yellow-headed blackbirds and pied-billed grebes occasionally nest here. A king rail was seen here in 1992. The best viewing is from the road on the south side of the wetland. Common yellow-throats, yellow warblers and yellow-rumped warblers can be seen as well. Extensive renovation was completed in early 2004 followed by heavy shorebird usage in the spring and great waterfowl use in the fall. (Courtesy Eric Volden)

21. York County (Map 29)

York County is a Rainwater Basin county with nearly 3,000 acres of permanent surface water, 2,400 acres of wooded habitats, and over 50,000 acres of grasslands or farmlands. There are tourist accommodations at Henderson and York.

A. Federal Areas

1. County Line Marsh WPA (Map location 8). Area 232 acres wetland, 176 acres upland. The county road leads to this marsh, which usually floods the road in spring. Usually large flocks of dabbling ducks gather here in early March. *(See also* Fillmore County)

2. Waco Basin WPA (Map location 3). Area 159 acres. Located 0.75 miles west and a ½ mile north of Waco in York County. Waco WPA has 113 acres of mixed marsh habitat and 46 acres of upland habitat. Recent renovation work makes the future look good for this site. It could become one of the better overall sites in the region. Recently a large water concentration pit was filled by putting the same soil back into the pit. With the pit restoration, much of the reed canarygrass, which

had been choking out all the other wetland plants, was dug up and deposited in the pit. The wetland basin is now deeper, making it very difficult for canarygrass to re-establish. Waco Basin Waterfowl Production Area adjoins Spikerush WMA, and has a 15-acre lake stocked with fish. (Courtesy Eric Volden)

3. Sinninger Lagoon WPA (Map location 7). Area 37 wetland acres, 123 upland acres. Located 2.5 miles east and 2 miles south of McCool Junction in York County. This site includes 42 acres of wetland and 118 acres upland. The cattle yard basin is best viewed in the evening with tough viewing in the morning sun. Satellite basins such as South Sinninger (southeast edge of WPA) and Q2 Basin (intersection of county roads Q and 2) can also be very productive if in the neighborhood. Between 1997 and 2001 this marsh supported over 6,000 spring shorebirds. Maybe the best all-around basin for ease of viewing, number and variety of birds. Expect good numbers of Hudsonian godwit and ruddy turnstone. Also very good in the fall with area highs for willet, long-billed dowitcher, red knot and Hudsonian godwit. (Courtesy Eric Volden)

4. Krause Lagoon WPA. 527 acres. Located 4.5 miles north of Shickley. Habitat includes 303 wetland acres and 224 upland acres. Heavy growth of cattails. Very large area of native grasses surrounding the wetland, especially on the east. (Courtesy Eric Volden)

5. Tamora Basin WPA. 260 acres. Located in Seward County, 6 miles west and 2 miles south of Seward, this is the farthest east of the rainwater basins. This site includes 228 acres of wetlands and 52 acres of upland habitat. Recently renovated, it could be very good in the future. (Courtesy Eric Volden)

B. State Areas
1. Spikerush WMA (Map location 4). Area 194 acres. Consists of mixed marsh and upland habitats.

2. Kirkpatrick Basin WMA. (Map location 5 & 6). Area 615 acres. Contains 70 acres of semi-permanent wetlands, 175 acres of seasonal wetlands. The rest of the north area consists of upland grasses, and the southern WMA consists of a shallow

wetland of 305 acres. This is an excellent area in spring for seeing migrating ducks and geese, especially snow geese, and slightly later it attracts a host of shorebirds, including American avocets and long-billed dowitchers. The north area is visible from I-80.

3. Hidden Marsh WMA (Not shown on map, located 2 miles east of Spikerush WMA). Area 120 acres.

4. Renquist Basin WMA. Area 107 acres. Consists of mixed upland and marshland.

5. Kirkpatrick Basin North WMA. 356 acres. Located 3.5 miles west and 2 miles south of York. Take I-80 Exit 348 north a half mile and then go east 1 mile. The 356 acre north basin contains 70 acres of semi-permanent wetlands, 175 acres of seasonal wetlands, and the rest upland grasses. This is an excellent area in spring for viewing thousands of migrating ducks and geese, especially snow geese. April through June it attracts a host of shorebirds including American avocets and long-billed dowitchers. (Courtesy Eric Volden)

6. Kirkpatrick Basin South WMA. Located 4 miles east and 1.25 miles north of Henderson. Purchased in 1982 by the Nebraska Game and Parks Commission, the southern unit consists of a shallow 305-acre wetland. This is an excellent area in spring for seeing thousands of migrating ducks and geese, especially snow geese. April through June it attracts a host of shorebirds including American avocets and long-billed dowitchers. (Courtesy Eric Volden)

C. Other Areas
1. Recharge Lake NRD Recreation Area. Area 120 acres, with 50-acre reservoir. Located 1.5 miles west of York. Developed for fishing, there are also hiking trails. Yellow-crowned night heron, piping plover, black tern, western sandpiper and dunlin are a few species recently reported from this site. (Courtesy Eric Volden)

22. Kearney County (Map 30)

Kearney County is a Platte Valley county with about 200 acres of permanent surface water, 300 acres of wooded habitats, and over 70,000 acres of grasslands or farmlands. Tourist accommodations are at Minden.

A. Federal Areas (all the following sites are temporary wetlands of fairly small size, but might be attractive to migrant water birds during wet springs).

1. Bluestem Basin WPA (Map location 4). Area 44 wetland acres, 32 upland acres.

2. Gleason Lagoon WPA (Map location 5). Area 197 wetland acres, 372 upland acres. Located 4 miles south and 4 miles west of Minden. Good waterfowl and shorebird viewing during spring migration depending upon water conditions. Habitat includes 195 acres of wetland and 372 acres of upland. Recent Management has attempted to control the invasive vegetation growth. Water can be pumped during dry migratory seasons. It offers a good variety of waterfowl, waders, and shorebirds. White-faced ibis, pectoral sandpipers, American bittern, black-crowned night herons, and whooping cranes have been reported here in recent years. (Courtesy Erick Volden)

3. Prairie Dog Marsh WPA (Map location 6). Area 430 acres, wetland, 382 upland. Located 5.5 miles south of Axtel. Habitat includes 471 acres of wetland and 421 acres of upland. A small black-tailed prairie dog colony exists on the higher ground near the southeast end of the WPA and is consistently used by burrowing owls. Historically known for its waterfowl concentrations. Dry years, land-leveling, irrigation reuse pits and invasive vegetation have reduced its attractiveness. Extensive work was done on this site in 2002 including rebuilding a dike going through the wetland so at least half of the wetland would have water during pumping, and filling of two small pits on the eastern section which were allowing water to seep out of this wetland rapidly. Fall use by ducks, especially blue-winged teal and pintails, was tremendous in 2002. Whooping cranes have been observed here in April and it is a great place for waders and shorebirds in late spring and late summer. Best viewing from the south parking area and may require some walking to get a good view of the birds. (Courtesy Eric Volden)

4. Lindau Lagoon WPA (Map location 7). Area 105 acres wetland, 47 upland acres.

5. Clark Lagoon WPA (Map location 12). Area 227 wetland acres, 222 acres upland.

6. Youngson Lagoon WPA (Map location 8). Area 113 acres wetland, 70 acres upland. Located 6 miles south and 0.5 miles east of Norman. Habitat includes 113 acres of wetland and 70 acres of upland. At times good to excellent for shorebirds and waterfowl. These two areas are only a couple of miles apart and are surrounded by areas with gullies and outpost sandhills. (Courtesy Eric Volden).

7. Frerichs Lagoon WPA (Map location 10). Area 33 acres wetland, 10 acres upland.

8. Killdeer Basin WPA (Map location 11). Area 36 acres wetland, 2 acres upland.

9. Jensen Lagoon WPA (Map location 9). Area 187 acres wetland, 278 acres upland.

B. State Areas
1. Hike-Bike bridge (Map location 1). This is a very good area for watching sandhill cranes at sunset and sunrise. Sometimes American woodcocks can be seen displaying near the north end of the bridge at sunset. Stop at the Fort Kearney State Historical Park for information and a park permit. The four-mile trail leads to Bassway Strip WMA along the two northernmost channels of the Platte (see Buffalo County).

2. Fort Kearny SRA (Map location 2). Area 163 acres. This area has primitive camping facilities and provides nearby parking for the hike-bike bridge.

3. Fort Kearny State Historical Park (Map location 3). Located 3 miles south and 4 miles west of I-80 exit 272 at Kearney. The park has a restored version of Fort Kearny including a stockade, parade grounds, blacksmith shop, and pony express stage station. The Fort was originally built to protect Overland Trail travelers. It is also a place from which one can watch field-feeding sandhill cranes in the spring. This area has an interpretive center and concessions area and offers primitive camping. There is a nice bird-feeding station located by the rangers' quarters north of the interpretive center. The historical park blends the history of the Platte River Valley with its ecology and natural history. The Visitor Center opens in early March for the

crane migration and provides information about crane and waterfowl viewing. There is a gift shop and a variety of displays. The Hike/Bike Trail is a mile east of the Fort and is a well maintained handicapped-accessible trail across the Platte River on a former railroad bridge. It provides an excellent view of the river and the woods along its banks and islands. Birds that can be seen along the trail include bald eagles, geese and ducks in January and February; Sandhill cranes and American woodcock in March; warblers and other passerines in April, May and June. This is one of the few public state areas where hunting is not allowed so there are birds there in the fall. Arrangements for blind visits can be made here (see introduction to region). Park entry permit required. Phone 308/865-5305 for information. (Courtesy Eric Volden)
http://outdoornebraska.ne.gov/parks/guides/park-search/showpark.asp?Area_No=97

4. Northeast Sacramento WMA (Map location 7). Area 30 wetland acres, ten acres upland.

23. Adams County (Map 31)

Adams County is a Rainwater Basin county with about 900 acres of permanent surface water, 1,200 acres of wooded habitats, and over 83,000 acres of grasslands or farmlands. Tourist accommodations are at Hastings.

A. Federal Areas
  1. Weseman WPA. 160 acres. Located 9 miles west and 4 miles south of Hastings. Habitat includes 80 acres of wetland and 80 acres of upland. In the mid-1990s this property was purchased by The Nature Conservancy through a cooperative effort with the U.S. Fish and Wildlife Service and the Rainwater Basin Joint Venture. Restoration work included filling a large pit along the property's south boundary and reseeding the uplands to native grasses. A parking lot was installed on the east side. This wetland generally has standing water only after heavy precipitation events or snow melts. (Courtesy Eric Volden)

  2. Kenesaw Lagoon (Map location 1). 231 acres. Located 1 mile east and a ½ mile south of Kenesaw. This Federal Wildlife Production Area contains 161 acres of wetland and 70 acres of upland. It attracts a large variety of water birds during spring.

Birds are best observed from county roads on the south or west side of the lagoon. A small great blue heron colony once existed on the east side of the lagoon. The mudflat area on the southwest side attracts shorebirds, waterfowl, and waders in spring. Eight whooping cranes visited this site for a few days in April of 1994 and there were unconfirmed reports of whooping cranes using this WPA again in the fall of 1999. The property had been left idle for about 20 years before being purchased by the U.S. Fish and Wildlife Service in 1997. It underwent a restoration in 1999 which included removal of trees, filling pits, removing dikes, filling in ditches, building parking lots, and fencing the property so the area could be grazed in future years. Waterfowl use has been high since completing the restoration. (Courtesy Eric Volden)

B. State Areas
  1. Prairie Lake Public Use Area (Map location 3). Area 125 acres, 30-acre lake. Mainly a fishing lake, with limited attractiveness to birds.

  2. Crystal Lake SRA (Map location 4). Area 33 acres. Located in Adams County, 1.5 miles north of Ayr. Free camping and primitive facilities. Picnic shelters and electrical hookups are available. A Nebraska State Park entry permit is required. Surrounding mature woodlands are good for warblers, thrushes, sparrows and flycatchers. (Courtesy Eric Volden) The lake is attractive to a variety of waders and waterfowl. Mostly developed for fishing; state park entry permit required.
http://outdoornebraska.ne.gov/parks/guides/park-search/showpark.asp?Area_No=56

  3. DLD SRA (Map location 6). Area 7 acres. Primitive camping facilities; state park entry permit required.
http://outdoornebraska.ne.gov/parks/guides/park-search/showpark.asp?Area_No=61

C. Other Areas (see also Lingle, 1994)
  1. Little Blue River (Map location 2). The wooded riparian zone of this river should be searched for passerines during migration periods. The Little Blue River passes through the southern third of Adams County. The wooded riparian zone of this river contains cottonwoods and hackberries with a thick understory that should be searched for passerines during migration periods. It is the

dominant drainage feature of southern Adams County. The area attracts migrating passerines in the spring and fall. Early May and September are the best times for viewing birdlife (Courtesy Eric Volden).

2. Hastings Museum and Lake Hastings (Map location 7). The Hastings Museum has a notable exhibit area for a small-town museum, including a diorama with ten whooping cranes. It sells materials of interest to naturalists, has an IMAX theater and planetarium, and provides advice on local attractions. Phone 402/461-4629 for information. Lake Hastings is a city-owned lake that might seasonally attract some birds, and is a short distance north of the Museum on routes 281 and 34. The Adams County Visitors Bureau (402/461-2370) might also be of assistance.

3. Ayr Lake (Map location 5). This is a privately owned seasonal wetland that sometimes attracts good numbers of migrating water birds. Access is limited to the peripheral road. Most noted for shorebirds and wading birds and to a lesser extent, waterfowl. April and May are the best times to find American golden-plovers, American avocets, and many other species. (Courtesy Eric Volden)

4. Holstein Hills. Located 2 miles west of Holstein and 20 miles east of Minden. This hilly, mixed-grass prairie region is home to several leks of displaying prairie chickens from March into early May each spring. Also common are grasshopper sparrows, dickcissels, horned larks and a variety of raptors. Drive the side roads, but be aware that many roads in this area are non-maintained dirt roads that are impassable when wet. All land in this area is privately owned so stay in your car on the county roads for viewing. (Courtesy Eric Volden)

24. Clay County (Map 28)

Clay County is a Rainwater Basin county with over 4,000 acres of permanent surface water, 900 acres of wooded habitats, and nearly 76,000 acres of grasslands or farmlands. There are tourist accommodations at Clay Center and Sutton.

A. Federal Areas (All of these are Waterfowl Production Areas that vary greatly in size and in relative wetland permanence).

1. North Hultine (formerly Sandpiper) WPA (Map location 6). Area 226 acres wetland, 214 acres upland. One of the best sites for seeing migrating shorebirds in the region, especially during late March and April.

2. Hultine WPA (Map location 7). Area 164 acres wetland, 74 acres upland.

3. Harvard Marsh WPA (Map location 8). Area 760 acres wetland, 724 acres upland. A deep, permanent marsh that attracts tens of thousands of snow, Canada and greater white-fronted geese each March. Access from the east is via a narrow, often slippery road, but is better from the south, at least to the railroad tracks. Driving beyond is not recommended after rains. There is also a parking area on the north side, but it is located quite far from the nearest water or marshy areas. Later on in spring this area is used by many shorebirds, including several sandpipers and piping plovers, and breeders include northern harriers and short-eared owls. Occasional flocks of sandhill cranes stop, and eagles are regular in early spring. Altogether one of the best birding wetlands in the entire region; up to 500,000 waterfowl have been seen here at the peak of spring migration. Later on as water levels drop, the main basin and several smaller wetlands to the south offer excellent shorebird watching. One of the better fall migration wetlands. Harvard WPA is one of the Rainwater Basin playa lakes. (Courtesy Eric Volden)

4. Lange Lagoon WPA (Map location 10). Located 0.25 miles east and 2 miles south of Sutton. Habitat includes 59 acres of wetland and 101 acres of upland. It is a migrant trap worth checking during migration. The grove of elm trees takes a few minutes to check and can be a good indicator of fallout. A nice area of permanent water exists but often cannot be viewed because of the immense stand of cattails. View the wetland from the east side and check out the trees on the north side. (Courtesy Eric Volden)

5. Theesen Lagoon WPA. (Map location 11). Located in Clay County, 1.5 miles northwest of Glenville. Habitat includes 46 acres of wetlands and 34 acres of uplands . Very consistent and productive during the mid-1990s, has been disappointing in recent dry years. Private property across the road

to the west of Theesen WPA usually has good mudflats for shorebirds in spring and late summer. (Courtesy Eric Volden)

6. Massie Lagoon WPA. (Map location 12). Located in Clay County, 2.5 miles south of Clay Center. Habitat includes 494 acres of wetland and 359 acres of upland. Massie is an excellent basin to see shoreline, edge and grassland bird species and is one of the best Rainwater Basin lagoons for waterfowl and shorebirds. It is part of a large elliptical basin that once occupied almost 2 sections. Between 1997 and 2001 this marsh supported over 7,000 spring shorebirds. Waterfowl species include snow geese, greater white-fronted geese, Canada geese, pintails, and mallards. An observation blind is located close to the parking lot on the south side of the lagoon. One of the best of the Clay County lagoons for waterfowl and shorebirds. An observation blind is located close to the parking lot on the south side of the lagoon; this access point is recommended over the others. Water levels in spring are maintained by pumping. (Courtesy Eric Volden)

7. Glenvil Basin WPA. (Map location 13). Area 83 acres wetland, 37 acres upland.

8. Kissinger Basin WPA. (Map location 14). Area 342 acres.

9. Meadowlark WPA. (Map location 15). Area 45 wetland acres, 35 upland acres.

10. Harms WPA. (Map location 16). Area 34 wetland acres, 25 upland acres.

11. Moger WPA. (Map location 17). Located 3 miles east and 2 miles south of Clay Center. Habitat includes 72 acres of wetland and 123 acres of upland. Spring burning and grazing have opened it up. A new well and pipeline allow adding water beyond spring runoff levels. Heavy cattails, no trees and an abundance of prairie grasses are found at Moger WPA. (Courtesy Eric Volden)

12. Shuck WPA. (Map location 18). Area 56 acres wetland, 24 upland acres.

13. Green Acres WPA. (Map location 19). Located 6 miles east and 4 miles south of Clay Center. Habitat includes 48 acres of wetland and 15 acres

of upland. In 1999 about 44 acres of wetland was disked in late September. A monotypic stand of river bulrush existed. The wetland remained dry through the fall and winter. In late April, 2000, a prescribed burn was conducted. Most of the organic matter and exposed tubers were burned. By October, about 50 percent of the wetland was again covered with bulrush. The remaining area contained smartweed. (Courtesy Eric Volden)

14. Eckhardt Lagoon WPA. (Map location 20). Area 66 acres wetland, 108 acres upland. Located 8 miles east and 4 miles south of Clay Center. Eckhardt WPA recently underwent a prescribed burn followed by disking the wetland and then grazing to reduce the dense stand of river bulrush and to keep vegetation short and in isolated pockets of the wetland. The wetland was pumped in the spring of 2000 and received tremendous use by pintails, mallards, white-fronted geese, and Canada geese in February and March. Near the end of March, snow geese began using the wetland as well. Located near Green Acres Waterfowl Production Area. (Courtesy Eric Volden)

15. Smith Lagoon WPA. (Map location 21). Located 6 miles south and 3.5 miles east of Clay Center. Habitat includes 226 acres of wetland and 254 acres of upland. In 1998, a new boat ramp was built in the southwest corner of the property. The west half was burned in the spring of 1999. Disking and grazing were done to open small pockets of water for pintail and mallard use. The small open-water pockets are not attractive to snow geese, which use the larger, more open wetlands. During the spring of 2000 migration, the wetland was heavily used by pintail and mallard. Snow geese begin using the wetland towards the end of their migration. One of the reasons for their using the wetland may have been hunting pressure on larger open wetlands. The dry conditions of 2002 and 2003 allowed for disking most of the bulrush in the wetland. An excellent waterfowl area during spring migration. (Courtesy Eric Volden)

16. Greenhead WPA. (Map location 22). Area 60 acres. Includes a dugout pond and mainly marshy habitats.

17. Hansen Lagoon WPA. (Map location 23). Located in Clay County, 0.25 miles west and 3.5 miles

north of Ong. Habitat includes 147 acres of wetland and 173 acres of upland. Excellent basin for bird-watching, the east portion has been best in recent years. Overgrown with cattail, one may need to walk from parking area west to large lagoon for best viewing. (Courtesy Eric Volden)

B. State Areas

1. McMurtry Refuge (Map location 9). Area 1,071 acres. No public access, and closed to hunting.

2. Bluewing WMA. Located four miles west, 0.5 miles south of Edgar. Includes 160 acres of lowland and seasonal wetland habitat.

3. Bullrush WMA. Located three miles west of Edgar. Includes 160 acres of upland and marshes.

4. Greenwing WMA. (Map location 24). Area 80 acres. Includes marsh, uplands, and scattered thickets. Located 0.5 miles east and 3 miles north of Ong. Habitat includes 53 acres of marsh and 27 acres of uplands with scattered thickets and cropland. It was purchased by the Nebraska Game and Park Commission in 1982 and had a restoration project completed in 2000, which included filling concentration pits and removing trees. (Courtesy Eric Volden)

5. Whitefront WMA. Located 1.5 miles west and 1.5 miles north of Clay Center. Includes 7 acres of permanent wetland, 158 acres of cropland and 10 acres of pasture.

25. Fillmore County (Map 29)

Fillmore County is a Rainwater Basin county with 1,600 acres of permanent surface water, 2,100 acres of wooded habitats, and about 55,000 acres of grasslands or farmlands. Tourist accommodations are at Geneva.

A. Federal Areas

1. County Line Marsh WPA (Map location 8). Area 408 acres. Located 3 miles south and 3 miles east of McCool Junction. This site contains 224 acres of wetland and 182 acres of upland. The county road leading to this marsh sometimes floods in spring. Very large flocks of dabbling ducks gather here in early March. Recent Management: The first tract purchased for this WPA was in 1964. The property has some challenges when it comes

to management. The wetland stays wet enough to support a very dense stand of bulrush leaving only a small amount of open water available for migratory waterfowl. The property is also located with half in York county and the other half in Fillmore. Numerous modifications in the hydrology have also occurred, including pits and drainage ditches. In 2002, conditions were dry enough to burn off the upland and a large portion of the wetland and disk it in late summer. No well exists on this property. However, some pumping was done in the fall of 2002 through an arrangement with the neighboring landowner. (Courtesy Eric Volden)

2. Real WPA (Map location 9). Area 121 wetland acres, 39 upland acres.

3. Bluebill WPA (Map location 10 right) & Marsh Hawk WPA (Map location 10 left). Areas 60 acres and 173 acres. Bluebill includes two marshes separated by higher ground. Marsh Hawk is mostly comprised of seasonal wetlands, with some trees and shrubs.

4. Wilkins Lagoon WPA (Map location 11). Area 370 acres wetland, 160 acres upland. Located 1 mile east and 2 miles south of Grafton. Habitat includes 321 acres of wetland and 208 acres of upland habitat. Another large "spring" basin that should be checked during that particular season. Dirt roads from every direction can make access impossible during wet spells. (Courtesy Eric Volden)

5. Morphy Lagoon WPA (Map location 12). Area 76 wetland acres, 12 upland acres.

6. Rolland Lagoon WPA (Map location 13). Area 53 wetland acres, 76 upland acres.

7. Rauscher Lagoon WPA (Map location 14). Area 140 acres wetland, 111 acres upland.

8. Weiss Lagoon WPA (Map location 18). Area 40 upland acres, 120 wetland acres.

9. Krause Lagoon WPA. (Map location 16). Area 303 wetland acres, 224 upland acres.

10. Mallard Haven WPA (Map location 17). Located 2 miles north of Shickley. Habitat includes 633 acres of wetland and 454 acres of upland. One

of the largest and perhaps best marshes for waterfowl during spring. This basin provides habitat for thousands of white-fronted geese, snow geese and ducks during late February and early March. Many wetland birds remain here to breed, including northern harriers, great-tailed grackles, and yellow-headed blackbirds. Between 1997 and 2001 this marsh supported over 8,000 spring shorebirds. Best in spring, sometimes a dud but can be very good. There are several parking areas around the site's perimeter and a kiosk with information in the southeast parking lot. Intense disking and burning coupled with grazing and pumping over the last few years have helped to decrease the dense stands of vegetation. (Courtesy Eric Volden). There are several parking lots at access points.

11. Brauning WPA. 240 acres. 3 miles south and 3.5 miles west of Grafton. Includes 165 acres of wetlands and 75 acres of uplands.

B. State Areas:
1 Sandpiper WMA (Map location 15). Area 160 acres. Includes 56 acres of marsh, with plum, willow, cottonwood and osage orange also present. Located 5 miles west and 1.5 miles south of Geneva. Habitat includes 226 acres of wetland and 214 acres of upland with plum, willow, cottonwood and osage orange present. Purchased in 1984 by NGPC and restoration was completed in 2001 including sediment and tree removal. One of the best sites for seeing migrating shorebirds in the region, especially during late March and April. (Courtesy Eric Volden)

26. Franklin County (Map 30)

Franklin County is a Republican Valley county with about 1,500 acres of surface water, over 5,000 acres of wooded habitats, and over 160,000 acres of grasslands or farmlands. There are no tourist accommodations in the county.

A. Federal Areas
1. Ritterbush Marsh WPA (Map location 13). Area 49 acres wetland, 32 acres upland.

2. Quadhammer Marsh WPA (Map location 14). Area 308 wetland acres, 286 acres upland.

3. Macon Lakes WPA (Map location 15). Area 498 acres wetland, 466 acres upland.

B. State Areas
1. Ash Grove WMA. Area 74 acres. Includes rolling hills, grasses, rock outcrops, and a small stream.

2. Limestone Bluffs WMA. Area 479 acres. Includes rolling hills with grasses, rock outcrops, and wooded ravines with a spring-fed stream.

27. Webster County

Webster County is a Republican Valley county with over 2,600 acres of surface water, nearly 4,000 acres of wooded habitats, and over 160,000 acres of grasslands or farmlands. Tourist accommodations are at Red Cloud.

A. Federal Areas: None

B. State Areas
1. Elm Creek WMA. Area 120 acres. Located three miles south of Cowles. Mostly wooded, with a creek and slough at one end.

2. Indian Creek WMA. Area 114 acres. Located one mile south of Red Cloud, containing riparian woods along the Republican River. Wood ducks, ospreys and eagles might be seen, plus wooded habitats woodpeckers and passerines.

C. Other Areas.
1. Willa Cather Memorial Prairie. 610 acres. A Willa Cather Foundation loess hills prairie, located six miles south of Red Cloud. It is grazed but unplowed mixed-grass prairie. From 164 to 219 plant species have been found in remnant loess hills prairies of this region.

28. Nuckolls County

Nuckolls County is a Republican Valley county with 1,200 acres of surface water, over 10,000 acres of wooded habitats, and over 125,000 acres of grasslands or farmlands. There are tourist accommodations in Nelson and Superior.

A. Federal Areas: None.

B. State Areas

1. Smartweed Marsh WMA. Area 74 acres wetland, 6 acres upland. Located two miles south and two miles west of Edgar. Mostly grassy lowlands, but with some marshy areas.

2. Smartweed Marsh West WMA. Area 38 acres. Located one mile south and three miles west of Edgar. Mostly grassy lowlands, but with some upland habitats.

29. Thayer County

Thayer County is a dissected plains county with about 1,800 acres of surface water, over 5,000 acres of wooded habitats, and 113,000 acres of grasslands or farmlands. Tourist accommodations at Hebron.

A. Federal Areas: None

B. State Areas

1. Little Blue WMA. Area 303 acres. Located three miles east of Hebron. Mostly flat, wooded bottomland of the Little Blue River, with some grasslands and croplands,

2. Prairie Marsh WMA. Area 160 acres. Located two miles west of Bruning. Consists of seasonal wetlands and adjoining uplands.

3. Meridian WMA. 190 acres, with virgin tall-grass and mixed-grass prairie, 3.5 miles north of Gilead. (402/749-7650).

Tall-grass Prairie Sites in East-Central Nebraska

Birders who wish to find tallgrass prairie birds in east-central Nebraska should consider visiting these sites:

Antelope County (p. 43)
Grove Lake WMA. Some sand and gravel prairie 100 yards northeast of parking area. See text description.

Boone County (p. 45)
Olson Nature Preserve. 112 acres of sandhill prairie and oak woodlands, 9 miles northwest of Albion, and 1 mile west. Property of Prairie Plains Resource Institute (402/386-5540 or 402/694- 5535).

Buffalo County (p. 46)
Lillian Annette Rowe Sanctuary & Iain Nicolson Audubon Center. About 420 acres of native prairie, plus 220 acres of restored prairie.

Hall County (p. 48)
The Crane Trust. Contains 6,400 acres of native and re-seeded wet meadows. See text description.

Hamilton County (p. 51)
Lincoln Creek Prairie and Trail, 16-acre loop trail on Lincoln Creek, east side of Aurora. Property of Prairie Plains Resource Institute (402/386-5540 or 402/694-5535).

Knox County (p. 43)
Bohemia Prairie WMA. Nearly 600 acres of virgin prairie. See text description.
Greenvale WMA. 70 acres of virgin prairie among woodlands. On Verdigre Creek, 9 miles west, 3 miles south of Verdigre.
Niobrara State Park. See text description. Ask Park Office for prairie location.

Madison County (p. 45)
Oak Valley WMA. Mixed virgin prairie and oak draws. See text description.

Merrick County (p. 50)
Bader Park Natural Area. Lowland prairie, riparian woodlands. See text description.

Nance County (p. 46)
Sunny Hollow WMA. 160 acres of wet virgin prairie and wetlands. See text description.

Platte County (p. 45)
Wilkinson WMA. 80 acres of wet virgin prairie. See text description.

Thayer County (p. 59)
Meridian WMA. 190 acres, tallgrass and mixed-grass prairie. See text description.

Sandhill Cranes

The Eastern Region: Lower Platte and Missouri Valleys

This region is a land that once was ruthlessly scraped over by glaciers, and which later was mantled by tall-grass prairies and riparian deciduous forests with eastern biogeographic affinities. It is bounded to the east by the Missouri River, which is now mostly channeled and much degraded as far as wildlife habitat is concerned. However, some stretches, such as around Ponca State Park, provide a faint idea of what the river once was like. The Missouri Valley is still a migratory pathway not only for arctic-breeding waterfowl such as snow geese, which alone now number over a million birds using this narrow flyway, but also myriads of forest-adapted Neotropic migrants, especially warblers and vireos. Remnant stands of mature deciduous forest still exist at Rulo Bluffs Preserve and Indian Cave State Park in Richardson County, Fontenelle Forest and Neale Woods in the Omaha area, DeSoto National Wildlife Refuge near Blair, and Ponca State Park in Dixon County. These are among the best places that can be visited in early May to see such wonderful birds as they journey north to breeding grounds in the Upper Midwest and southern Canada.

The region is the most heavily populated part of the state, and thus has the fewest areas of native prairie vegetation persisting, but in such areas grassland species such as greater prairie-chickens still gather at sunrise every spring, on traditional display grounds made sacred by decades if not centuries of use. Similarly, other prairie species such as long-billed curlews and upland sandpiper rarely occur here any more, but grassland sparrows still announce their territories from fence posts, and house wrens, gray catbirds and brown thrashers sing from plum thickets along roadside ditches. Many other eastern or southeastern species occur and locally breed here, including Henslow's sparrow, Kentucky, northern parula, and prothonotary warblers, and possibly red-shouldered and broad-winged hawks. Further-

more, chuck-will's-widows certainly must nest in the wooded habitats bordering the southeastern corner of the state, and pileated woodpeckers sometimes also nest here, making it an area of special interest to birders.

Nebraska breeders that are largely limited to the forested Missouri Valley are the American woodcock, barred owl, chuck-will's widow, whip-poorwill, ruby-throated hummingbird, yellow-throated vireo, tufted titmouse, blue-gray gnatcatcher, Louisiana waterthrush, Kentucky warbler, summer tanager and scarlet tanager. Cerulean warblers occur very locally. More than 100 bird species are believed to nest in Iowa's loess hills region on the east side of the Missouri River, including Iowa's densest breeding populations of the turkey vulture, American kestrel, Bell's vireo, orchard oriole, chuck-will's-widow, and summer tanager. The most common nesting species in the Loess Hills region include the brown-headed cowbird, northern cardinal, brown thrasher, house wren, mourning dove, American crow, blue jay and red-headed woodpecker.

1. Cedar County

Cedar County is a Missouri Valley county with 3,900 acres of surface water, 10,700 acres of wooded habitats, and almost 134,000 acres of grasslands or farmlands. There are tourist accommodations at Hartington, Laurel and Randolph.

A. Federal Areas.
1. Gavin's Point Dam. See state highway map for location. Birding from the dam should offer views of gulls, waterfowl, and other birds including numerous bald eagles during migration periods. A nature trail is present, as well as an aquarium, in the associated Lewis and Clark Recreation Area. www.nwo.usace.army.mil/html/Lake_Proj/gavinspoint/

2. Lake Yankton. Located just below Gavin's Point Dam; partly in South Dakota. A relatively small reservoir below the spillway at Gavin's Point Dam that offers good birding along its wooded shoreline.

B. State Areas

1. Chalkrock WMA. Area 130 acres. Located 4 miles south of the Missouri River bridge on Highway 81 and 1.5 miles west of the highway. Consists of 90 upland acres and a 45-acre, reservoir.

2. Wiseman WMA. Area 365 acres. Located 1 mile north and 5 miles east of Wynot. Just south of the Missouri River, this area includes steep wooded bluffs, and grassy ridges. The woods are mostly bur oak, cedar, hackberry and ash.

2. Dixon County (Map 32)

Dixon County is a Missouri Valley county with over 5,000 acres of surface water, 10,000 acres of wooded habitats, and over 73,000 acres of grasslands or farmlands. There are tourist accommodations at Ponca.

A. Federal Areas: None

B. State Areas

1. Ponca State Park. (Map location 1). Area 2,166 acres. This park is mostly forested with stands of bur oak, walnut, hackberry and elms; one of the oaks is more than 300 years old. A bird list of 297 species and 70 breeders covers the park plus adjoining parts of north Nebraska, southeastern South Dakota and northwestern Iowa. There are 17 miles of hiking trails and modern cabins are available, as is an undeveloped campground. Whip-poor-wills are common in summer, and bald eagles are present during much of the year. Snow geese migrate past the area in spring and fall, and the nearby Missouri River is still unchanneled here, thus it resembles its original state. Classified as a Nebraska Important Bird Area. State park entry permit required.
http://outdoornebraska.ne.gov/parks/guides/park-search/showpark.asp?Area_No=143

2. Buckskin Hills WMA. Located 2 miles west and 2 miles south of Newcastle. Consists of 340 acres of grasslands and woods around a 75-acre reservoir.

3. Elk Point Bend WMA/SRA. Under development. Located about 2 miles north of Ponca State Park. Oak savanna and riparian wetlands near the Missouri River. Excellent for waterfowl watching during migrations. Not located on map; for information contact Ponca State Park (402/755-2284).

3. Dakota County (Map 32)

Dakota County is a Missouri Valley county with 3,400 acres of surface water, 6,800 acres of wooded habitats, and over 28,000 acres of grasslands or farmlands. Tourist accommodations at South Sioux City.

A. Federal Areas: None.

B. State Areas

1. Basswood Ridge WMA (Map location 2). Area 360 acres. Consists of very rugged and heavily wooded uplands, with some Native American petroglyphs near the north end.

2. Omadi Bend WMA (Map location 3). Area 33 acres. Consists of bottomland forest along an oxbow lake.

4. Wayne County

Wayne County is a county of glaciated uplands, with 180 acres of surface water, 2,100 acres of wooded habitats, and nearly 46,000 acres of grasslands or farmlands. There are tourist accommodations at Wayne.

A. Federal Areas: None

B. State Areas

1. Sioux Strip WMA. Area 25 acres. Located at western edge of Scholes. Consists of upland grasses along an old railroad bed.

5. Thurston County

Thurston County is a Missouri Valley county with nearly 1,500 acres of surface water, over 22,000 acres of wooded habitats, and nearly 34,000 acres of grasslands or farmlands. There are no tourist accommodations in the county.

A. Federal Areas
1. Missouri River Federal Access Areas. Numerous access points for boats, as posted.

B. State Areas: None

6. Stanton County

Stanton County is an Elkhorn Valley county with less than 800 acres of surface water, 5,200 acres of wooded habitats, and 75,000 acres of grasslands or farmlands. There are no tourist accommodations in the county.

A. Federal Areas: None.

B. State Areas
1. Red Fox WMA. Area 537 acres. Located 1 mile south of Pilger. Includes a wooded remnant oxbow, a 25-acre sandpit lake, 0.6 miles of Elkhorn River frontage, and 163 acres of grassland .

2. Wood Duck WMA. Area 1,528 acres. Located about 2 miles south and 4 miles west of Stanton. Consists of riparian wooded habitats bordering the Elkhorn River, with several oxbow lakes and a stream. Many eastern songbirds nest here, and the marshy lakes are used by large numbers of geese, ducks, pelicans, cormorants and occasional swans. The perimeter roads are often wet and sometimes even flooded during wet springs, so caution is needed when driving.

Other Areas
1. Maskenthine Reservoir. An NRD flood-control reservoir (98 acres) located 1 mile north of Stanton. Open water, wetland, riparian woodland and a trail around the lake. Not shown on map.

7. Cuming County

Cuming County is an Elkhorn Valley county, with about 800 acres of surface water, 3,000 acres of wooded habitats, and over 75,000 acres of grasslands or farmlands. There are tourist accommodations at Beemer, West Point and Wisner.

A. Federal Areas: None

B. State Areas

1. Black Island WMA. Area 240 acres. Located 1 mile north and 4 miles west of Wisner. This area is a mixture of woods and grassland along 0.75 miles of the Elkhorn River. From the intersection of highways 275 and 15 (in Stanton County), drive 1 mile east, 0.5 miles south, 1.5 miles east, then south on entrance road. A mixture of woods and grassland along 0.75 miles of the Elkhorn River. "This area consists of wooded habitats (primarily cottonwoods), grassy/weedy vegetation, and some wet grassy areas. It also has access to the Elkhorn River. There are possibilities for migrating, breeding and wintering passerines as well as waterfowl, shorebirds and other water birds. This is not a 'go out of your way' type of area, but if making a trip along Highway 275, it might be a nice side stop. The area is open to public hunting and fishing, use caution during hunting seasons." (Information courtesy of T. J. Walker, via NOU website.)

C. Other Areas
1. Elkhorn River. This river is bounded by deciduous riverine forest along most of its length, and should provide for good birding opportunities.

2. West Point City Park. Municipal park, with some oxbow lakes and large trees. Not shown on map.

8. Burt County

Burt County is a Missouri Valley county with 3,200 acres of surface water, 5,500 acres of wooded habitats, and almost 39,000 acres of grasslands or farmlands. There are tourist accommodations at Lyons, Oakland and Tekamah.

A. Federal Areas: None

B. State Areas
1. Decatur Bend WMA. Area 133 acres. Located 3 miles east of Decatur.

2. Pelican Point SRA. Area 36 acres. Located along Missouri River, 3 miles north and 6 miles east of Tekama. Consists of riverine wooded habitats. State park entry permit required. "Pelican Point State Recreation Area is 4 miles east, 4 miles north and 1 mile east from Tekamah. This area includes a small, peaceful, primitive campground and access to the Missouri River. The campground

and surrounding area is dominated by large cottonwoods with some shrubby understory. This is a good place to see migratory and breeding bird species including warblers, vireos, thrushes, orioles, flycatchers, woodpeckers and other passerines as well as larger birds that tend to follow the river during migration." (Information courtesy of T. 1. Walker, via NOU website.)
http://outdoornebraska.ne.gov/parks/guides/parksearch/showpark.asp?Area_No=223

3. Summit Reservoir SRA. Area 535 acres. Located 3 miles west and 1 mile south of Tekama. The 190-acre reservoir is developed for fishing. State park entry permit required.
http://outdoornebraska.ne.gov/parks/guides/parksearch/showpark.asp?Area_No=247

9. Colfax County

Colfax County is a Platte Valley county with about 1,300 acres of surface water, 5,900 acres of wooded habitats, and 44,000 acres of grasslands or farmlands. There are tourist accommodations at Schuyler.

A. Federal Areas: None

B. State Areas
1. Whitetail WMA. Area 216 acres. Located 2 miles south and 1 mile west of Schuyler. Consists of 93 acres of Platte River bottomland forest and 123 acres of islands and river. "This Wildlife Management Area features a variety of wooded habitats ranging from cottonwood savannah to open woods with shrubs to dense cottonwood forest. There are some shallow oxbow wetlands and access to the Platte River. There is also a sandpit northwest of the parking lot (on private property) that occasionally has gulls, terns and waterfowl on it. A good area for migratory/breeding bird passerine species, shorebirds and larger migratory species. The area is open to public hunting and fishing, use caution during hunting seasons." (Information courtesy of T. J. Walker, via NOU website.)

10. Dodge County (Map 33)

Dodge County is an Elkhorn Valley county with 1,800 acres of surface water, over 3,000 acres of wooded habitats, and almost 40,000 acres of grasslands or farmlands. Tourist accommodations are at Fremont.

A. Federal Areas: None

B. State Areas
1. Dead Timber SRA (Map location 1, right). Area 150 acres, 50 acres lake. A recreational facility developed for fishing. An old oxbow lake, beside the Elkhorn River. The wooded high bank along the northeast part of the area has many seeps, holding many birds through the winter. State park entry permit required.
http://outdoornebraska.ne.gov/parks/guides/parksearch/showpark.asp?Area_No=59

2. Powder Horn WMA (Map location 1, left). Area 289 acres. Consists of riparian wooded habitats bounding the Elkhorn River, plus adjoining grasslands, marshes and croplands.

3. Fremont Lakes SRA (Map location 2). Area 670 acres. Includes 20 small sandpit lakes totaling 280 acres, with many recreational facilities. State park entry permit required.
http://outdoornebraska.ne.gov/parks/guides/parksearch/showpark.asp?Area_No=78

11. Washington County (Map 34)

Washington County is a Missouri Valley county with over 3,000 acres of surface water, nearly 15,000 acres of wooded habitats, and nearly 33,000 acres of grasslands or farmlands. Tourist accommodations are at Blair.

A. Federal Areas

1. DeSoto National Wildlife Refuge. (Map location 1). Total area 7,823 acres. This important national wildlife refuge is located around an old oxbow of the Missouri River, and consists mostly of riverine deciduous forest, an eight-mile oxbow lake, and croplands. It supports an enormous fall population of snow geese (and some Ross' geese), which may reach peak numbers of about 800,000 birds in late October or early November. There is a superb interpretive center, whose large windows face a 788-acre lake, allowing wonderful views of the geese, other waterfowl, and numerous bald eagles

in late fall and early spring. There are also outdoor viewing platforms for close viewing. There is a bird checklist of 240 species, including 81 breeders. Peak populations of ducks, mostly mallards, may reach 125,000. There is a twelve-mile drive around the refuge, and four hiking trails. A summary checklist of the refuge's birds is included in the supplement. The mailing address for the refuge is Rte. 1, Box 114, Missouri Valley, Iowa, 51555 (712/642-4121). The interior refuge roads are closed during some periods; inquire at the interpretive center (9 a.m. to 4 p.m. daily) for schedule. Daily admission fee charged, and seasonal driving restrictions. www.fws.gov/midwest/desoto

2. Boyer Chute National Wildlife Refuge (Map location 3). Located along the Missouri River, this new refuge of about 2,000 acres is a cooperative project involving the U.S. Corps of Engineers and several state agencies. There are currently several access points, with two miles of roads and about five miles of trails, including a mile of paved trail. The area is being managed from DeSoto National Wildlife Refuge, which should be contacted for information. Classified as a Nebraska Important Bird Area.
www.fws.gov/refuges/profiles/index. cfm?id=64640

B. State Areas
1. Fort Atkinson State Historic Park. (Map location 2). Probably mainly of interest for historical reasons, but no doubt migrating passerines use the area to some degree. Forested river bluffs and open grasslands, plus restored fort buildings. State park entry permit required.
http://outdoornebraska.ne.gov/parks/guides/park-search/showpark.asp?Area_No=73

12. Butler County

Butler County is a Platte Valley county with about 700 acres of surface water, 6,000 acres of wooded habitats, and 84,000 acres of grasslands or farmlands. Tourist accommodations are located at David City.

A. Federal Areas: None

B. State Areas

1. Redtail WMA. Area 320 acres, 17-acre reservoir. Located 1 mile east of Dwight. Includes grassland, wooded draws, and a pond.

2. Timber Point Public Use Area. Area 160 acres. Located 1 mile south and 2 miles east of Brainard. A 28-acre lake is present.

13. Saunders County (Map 35)

Saunders County is a Platte Valley county with over 4,000 acres of surface water, over 9,000 acres of wooded habitats, and 103,000 acres of grasslands or farmlands. There are tourist accommodations at Wahoo.

A. Federal Areas: None

B. State Areas
1. Jack Sinn WMA (Map location 7). See Lancaster County.

2. Red Cedar Public Use Area (Map location 2). Area 175 acres, with a 50-acre lake. Nearby Madigan Prairie (23 acres) is located 1 mile east of the Butler County line, 2 miles south of Route 92 (the eastern 23 acres of southern half of NW quadrant of Sec. 20, T 14 N, R 5 E.) Owned by the University of Nebraska Foundation; managed by the University of Nebraska School of Biological Sciences.

3. Larkspur WMA (Map location 3). Area 160 acres. Includes 37 acres of bur oak wooded habitats, plus areas of native prairie and seeded grassland.

4. Czechland Lake. An 85-acre lake. An NRD reservoir with grassland and riparian timber. Located 8 miles north and 3.5 miles west of Weston. Not shown on map.

14. Douglas County (Map 36)

Douglas County is a Missouri Valley county, with 12,800 acres of surface water, over 3,000 acres of wooded habitats, and 20,000 acres of grasslands or farmlands. There are tourist accommodations at Omaha.

A. Federal Areas: None

B. State Areas

1. Two Rivers SRA (Map location 1). SRA area 643 acres, WMA312 acres. The SRA is developed for recreational purposes. Located just south of the SRA, the WMA consists of timbered riverbottom forest, marshes, and croplands. State park entry permit required for the SRA.
http://outdoornebraska.ne.gov/parks/guides/park-search/showpark.asp?Area_No=175

C. Other Areas

1. Neale Woods Nature Center (Map location 7). Area 554 acres. This privately owned nature center includes nine miles of trails through hardwood forests, restored prairie uplands, and riverine wooded habitats. Open weekdays 8-5, and on Sunday afternoons. A checklist of 190 species either seen or expected in the area is available. There are 57 likely breeders, including barred owl, whip-poor-will, ruby-throated hummingbird, eastern wood-pewee, American redstart, scarlet tanager, rose-breasted grosbeak, and indigo bunting. The address is 14323 Edith Marie Ave., Omaha 68112 (402/453-5615). Classified as a Nebraska Important Bird Area. Admission fee.
www.fontenelleforest.org

2. Standing Bear Lake (Map location 5). Reservoir area 135 acres. A flood-control lake that attracts migrant waterfowl seasonally.

3. Glenn Cunningham Lake (Map location 6). Reservoir area 390 acres. A flood-control reservoir that is the largest in the Omaha area, and as such is important for migrating waterbirds.

4. Papio D-4 Lake. Located just east of Glenn Cunningham Lake. A 30-acre reservoir, which sometimes attracts migrant gulls and waterfowl.

5. Zorinsky Lake (Map location 9). A flood-control reservoir of about 250 acres.

15. Sarpy County (Map 36)

Sarpy County is a Missouri Valley county with 2,500 acres of surface water, nearly 15,000 acres of wooded habitats, and 21,000 acres of grasslands or farmlands. There are tourist accommodations at Bellevue, Gretna and Papillion.

A. Federal Areas: None

B. State Areas

1. Chalco Hills Recreation Area (Map location 2). Area 1200 acres; Wehrspan Reservoir 245 acres. This developed area has boating, an arboretum, a nature trail and natural resource center, a wildlife observation blind, and a 5-mile hike-bike trail.
www.papionrd.org/recreation_and_wildlife/chalco_hills.shtml

2. Schramm Park Recreation Area (Map location 3). Area 340 acres. This area includes the Ak-Sar-Ben aquarium, and some excellent wooded habitats that teem with warblers during spring migration. There are five miles of trails, and an educational center at the aquarium. Whip-poor-wills can be heard here, and Kentucky warblers probably breed. There is a three-mile loop trail leading from the parking area. State park entry permit required. Phone 402/332-3901 for information.
http://outdoornebraska.ne.gov/parks/guides/park-search/showpark.asp?Area_No=158

C. Other Areas

1. Fontenelle Forest Preserve (Map location 4). Area 1300 acres. This large area of mature riverine hardwood forest includes 17 miles of footpaths, as well as a mile-long boardwalk and a small combined nature center and museum. There is a bird checklist of 246 species that have been reported in the past decade, and more than 100 of these are summering species that potentially breed. The "Omaha" checklist in the supplement section is mostly based on checklists for Fontenelle Forest and Neale Woods. Summer species of special interest include American woodcock, broad-winged and red-shouldered hawks, whip-poor-will, Acadian flycatcher, Carolina wren, yellow-throated vireo, wood thrush, American redstart, cerulean, prothonotary and Kentucky warblers, brown creeper, and the scarlet and summer tanagers. The yellow-throated warbler has been reported to perhaps nest here too, a location well to the north of its known breeding range, and the pileated woodpecker has recently bred here. An observation blind overlooks a marsh, and there are organized bird or nature hikes, plus many other programs. The center is open 8-5 daily; admission fee. The address is Bellevue Blvd., Omaha 68005. Call 402/731-3140 for information.
www.fontenelleforest.org/come.html

2. Gifford Point (Map location 8). A 1,300-acre area of riverbottom forest, and a 400-acre farm (advance reservations needed for farm). Dedicated to environmental education, and located just east of Fontenelle Forest, at 700 Camp Road, Bellevue, NE 67005. Call 402/597-4920 for information.

3. Neale Woods Nature Center. See Douglas County.
www.fontenelleforest.org/come.html

16. Seward County (Maps 29, 37)

Seward County is a loess plains county with 1,500 acres of surface water, about 6,000 acres of wooded habitats, and over 110,000 acres of grasslands or farmlands. There are tourist accommodations at Seward.

A. Federal Areas: None

B. State Areas
1. Meadowlark NRD Recreation Area. (Map 37, location 1). Area 55 acres.

2. Oak Glen WMA (Map 37, location 2). Area 632 acres. Consists of 260 acres of mature oak wooded habitats, with some grassland including native prairie.

3. Branched Oak SRA (Map 37, location 3). See Lancaster County.
http://outdoornebraska.ne.gov/parks/guides/park-search/showpark.asp?Area_No=34

4. Bur Oak WMA (Map 37, location 4). Area 143 acres. Located along US 84, and comprised of mature bur oak wooded habitats, with some green ash and native grassland.

5. Twin Lakes WMA. (Map 37, location 5). Area 1,300 acres. Includes 255- and 50-acre reservoirs, plus marshes, wooded bottomlands, upland prairie, grasslands, and small ponds. Although part of the prairie has been re-seeded, other portions are of native prairie. Dickcissels, eastern and western meadowlarks, sedge wrens, eastern bluebirds, and other prairie or forest edge species occur here. The two lakes do not allow fishing with motors other than electric ones, and no waterfowl hunting is permitted, so undisturbed birding is possible. The

area is closed to public access from October 15 to the end of the hunting season for Canada geese.

6. North Lake Basin WMA (Map 29, location 2). Located 1 mile north of Utica. Contains 364 acres of marsh and adjoining uplands. Usually this area attracts large numbers of waterfowl during spring migration.

7. Freeman Lakes WMA (Map 29, location 1) Located northwest of Utica. Includes 42 upland acres and 146 acres of wetland. Similar to North Lake Basin.

17. Lancaster County (Maps 35, 38)

Lancaster County is in an area of glacial till plains, with nearly 6,000 acres of surface water, over 7,000 acres of wooded habitats, and 140,000 acres of grasslands or farmlands. Tourist accommodations are at Lincoln.

A. Federal Areas: None

B. State Areas: (Most of the following are flood-control reservoirs built in the 1960s to control flooding in the Salt Creek Valley. Silting-in has affected all of these reservoirs, producing marsh-like habitats at the places where creeks feed into the reservoirs, and such areas usually provide the best birding sites. Flooded trees are also usually present, providing perching sites for bald eagles and cormorants.

1. Pawnee Lake SRA (Map 38, location 4). Area 1906 acres, reservoir 740 acres. This fairly large reservoir (second to Branched Oak) attracts many migrant waterfowl during spring, and also has many prairie species.
http://outdoornebraska.ne.gov/parks/guides/park-search/showpark.asp?Area_No=135

2. Conastoga Lake SRA (Map 38, location 8). Area 486 acres, reservoir 230 acres. Similar to Pawnee Lake in its bird life, this lake is also surrounded by grasslands and some wooded habitats.

3. Yankee Hill Lake WMA . (Map 38, location 9). Area 728 acres. Includes a 210-acre reservoir surrounded with rolling grassland and wooded bottomland.
http://outdoornebraska.ne.gov/fishing/programs/lakemapping/district5.asp

4. Killdeer WMA (Map 38, location 10). Area 69 acres. Contains a 20-acre reservoir, with surrounding marsh, wooded draws, and uplands.

5. Bluestem Lake SRA (Map 38, location 11). Area 483 acres, reservoir 325 acres. The silted-in upper (northern) end of this reservoir is quite marsh-like, and is good for finding marsh birds. http://outdoornebraska.ne.gov/parks/guides/park-search/showpark.asp?Area_No=27

6. Olive Creek SRA (Map 38, location 12). Area 438 acres, reservoir 175 acres. This small lake seems to attract rare waterfowl, especially scoters, which turn up there almost every year. http://outdoornebraska.ne.gov/parks/guides/park-search/showpark.asp?Area_No=131

7. Teal WMA (Map 38, location 13). Area 66 acres. Contains a 27-acre reservoir, with surrounding wooded bottomland and rolling upland.

8. Stagecoach Lake SRA (Map 38, location 15). Area 412 acres, reservoir 120 acres. Located three miles north of Prague. http://outdoornebraska.ne.gov/parks/guides/park-search/showpark.asp?Area_No=170

9. Cottontail Public Use Area (Map 38, location 14). Area 148 acres. Includes a 28-acre reservoir.

10. Wagontrain Lake SRA (Map 38, location 17). Area 750 acres, reservoir 315 acres. Much like the other lakes in the area, with many flooded trees. Located three miles east of Hickman. http://outdoornebraska.ne.gov/parks/guides/park-search/showpark.asp?Area_No=180

11. Hedgefield Lake WMA (Map 38, location 18). Area 114 acres. Contains a 44-acre reservoir with surrounding rolling upland with some wooded vegetation.

12. Branched Oak Lake SRA (Map 35, location 5). Area 4,406 acres, reservoir 1,800 acres. This is the biggest of the reservoirs in the county, and one that seems to attract many rare birds (gulls, waterfowl, loons) during fall, winter and early spring. Vast flocks of snow geese visit it in early March, as well as Canada and greater white-fronted geese. Eagles are common during the spring when the

ice is breaking up, and ospreys may also be seen on migration. Many species of ducks and white pelicans are common during migration. The shorebirds are best during fall, and flooded timber at the northern end of the lake attracts cormorants too. Snowy owls and black-billed magpies often turn up here, and the brushy vegetation supports wintering American tree sparrows and Harris' sparrows among many others. State park entry permits are needed for some parts of the area. http://outdoornebraska.ne.gov/parks/guides/parksearch/showpark.asp?Area_No=34

13. Wildwood Lake WMA (Map 35, location 4). Area 491 acres. Includes a 107-acre reservoir surrounded by native wooded habitats, hilly grassland and crops. This is a little-visited lake, and an unusually beautiful one.

14. Jack Sinn Memorial WMA (Map 35, location 6 & 7). Area 632 acres. Consists of mostly seasonally wet lowlands that occur along a creek drainage, and include some beaver ponds up to six feet deep approximately. Some of the best rail and marsh bird habitat in the county occurs here; some of the best saline wetlands in eastern Nebraska are found here. There is an old railroad bed that provides dry walking opportunities in this usually wet environment.

15. Arbor Lake WMA (Map 38, location 1). Area 63 acres. Consists of a 63-acre saline and semi-permanent wetland that seasonally supports great-tailed grackles, migrant ducks and shorebirds, and prairie passerines.

C. Other Areas
1. Roper's Lake (undeveloped; Map 38, location 2). This little-visited part of Lincoln can only be reached by following 48th Street north to the city landfill, and stopping there for permission to go on to the marshes and riparian wooded habitats about a quarter mile beyond. It is well worth the effort; this area is perhaps the best place in Lincoln for seeing large numbers of species. Red-tailed hawks and great horned owls nest here, coyotes and deer are regularly seen, and the marshy areas attract rails, ducks, geese, cormorants, yellow-headed blackbirds, black-crowned night herons, and many other species. Care must be taken in walking through the area, which is partly land-

fill, and sometimes has ground irregularities that one can easily fall into or trip over. Ospreys and shorebirds are sometimes seen along Salt Creek, which borders the area on the north and west.

2. Nine Mile Prairie (Map 38, location 3). One of the few remaining native tall-grass prairies in the county, and a classic site of early ecological research by university personnel such as John Weaver. Only a small amount of water exists on the prairie, so the species are mostly upland passerines, such as grasshopper sparrows and dickcissels. The area occupies 240 acres, and is open year round. http://snr.unl.edu/aboutus/where/fieldsites/ninemileprairie.asp

3. Oak Lake Park (Map 38, location 5). Area 47 acres. A small park around two lakes that attract gulls, cormorants, geese, ducks, and other waterbirds during migration.

4. Capital Beach saline wetlands (Map 38, location 6). The site of a once-saline lake, the undeveloped east side still supports a good marsh habitat that is now being preserved, and has provided records of king rails and other uncommon to rare water-dependent birds. Enter from Sun Valley Boulevard onto Westgate Boulevard, then west on Lake Drive to parking area. Grassy wetland areas are excellent for sparrows. www.lpsnrd.org/Recreation/wetlands.htm

5. Pioneer's Park Nature Center (Map 38, location 7). Pioneer's Park is the oldest of Lincoln's city parks, and one of the largest (over 1,200 acres). It has a few small ponds, and some native prairie, but is mostly planted to pines and other conifers. These wooded habitats support great horned owls (year round) long-eared owls (in winter), red-tailed and sharp-shinned hawks, and many passerines. A nature trail extends out into prairie and brushland, and through riparian wooded habitats along a branch of Salt Creek, where wood ducks are common. A nature center offers views of wintering songbirds that gather near feeders, and looks out over a pond. A bird checklist has 237 species. Open Sundays noon to 5 p.m.; weekdays 8:30 a.m. to 5 p.m. or (summer) 8:30 p.m. Ph. 402/441-7895.

6. Holmes Lake (Map 38, location 16). Area 555 acres. A city park around a small reservoir that has a resident flock of Canada geese. In some winters ducks such as goldeneyes, scaup and sometimes hooded mergansers can be seen among the geese.

7. Wilderness Park (Map 38, location 14). Area 1455 acres. This long and narrow park follows the Salt Creek for about seven miles, and has about 20 miles of hiking and horseback trails. There are good stands of mature bur oak and riparian forest, and several species of owls are common (barred, eastern screech-owl and great horned owl), as well as several woodpeckers including red-bellied. It is probably the best Lincoln location for migrant warblers, and breeding songbirds include tufted titmouse (near its western limits), eastern bluebird, orchard and Baltimore oriole, Carolina wren (near its northwestern limits). A park checklist has 191 species.

8. Wyuka Cemetery. Not shown; entrance at 37th & O St., Lincoln. Another excellent location for migrant songbirds such as vireos and warblers, mainly along eastern edge.

9. Spring Creek Prairie Audubon Center (location 19.) 639 acres. Owner: National Audubon Society. Located three miles south of the west edge of Denton; follow 98th St. south out of Denton; entrance gate on east side of road. Some small wetlands including a spring, riparian wooded habitats, and hilly prairie uplands. Henslow's sparrows are regular here. Over 350 plant species have been recorded, and 210 species of birds. Monday-Friday 9 a.m.-5 p.m. Open on some weekends. Classified as a Nebraska Important Bird Area. Donation suggested. Ph. 402/797-2301. http://springcreekprairie.audubon.org/

10. Little Salt Creek Marsh. 280 acres. A Nature Conservancy wetland three miles west of Raymond, north side of Raymond Road. Saline wetlands similar to Arbor Lake. Ph. 402/342-0282.

18. Cass County (Map 39)

Cass County is a Missouri Valley county with about 1,800 acres of surface water, 24,000 acres of wooded habitats, and almost 56,000 acres of grasslands or farmlands. Tourist accommodations at Greenwood.

A. Federal Areas: None

B. State Areas

1. Eugene T. Mahoney State Park (Map location 1). Area 574 acres. This is a highly developed riverine woodland park, with lodging, cabins, eating facilities, and other popular attractions. It has an excellent population of eastern bluebirds, and borders the Platte River, where riparian deciduous forest is well developed. There is a tall observation tower built among a stand of bur oaks, and a 6-mile network of trails. State park entry permit required. Phone 402/944-2523 for information.
http://outdoornebraska.ne.gov/parks/guides/park-search/showpark.asp?Area_No=273

2. Platte River State Park (Map location 2). Another popular park, this area is quite similar to the last-named park, and also to Louisville Lakes State Park. Migrant warblers are abundant, and Kentucky warblers may breed here, as do scarlet and summer tanagers. There are two observation towers. Cabins are available; no camping is allowed. State park entry permit required. Phone 402/234-2217.
http://outdoornebraska.ne.gov/parks/guides/park-search/showpark.asp?Area_No=224

3. Louisville Lakes State Park (Map location 3). Area 142 acres. Much like the two previous parks, with both primitive and modern camping and various concessions. The nearby Schramm Park is less crowded and probably offers better birding. Requires state park entry permit.

4. Randall W. Schilling WMA (Map location 4). A 1,500-acre managed waterfowl area, with 25 acres of water and nearby cropland, mainly designed to attract snow geese. Open to the public from April 1 to September 30, and used for controlled-access goose hunting during the fall season.

5. Rakes Creek WMA (Map location 5). Located eight miles east and one mile south of Murray, 316 acres of upland. http://outdoornebraska.ne.gov/

19. Otoe County (Map 44)

Otoe County is a Missouri Valley county with 2,500 acres of surface water, over 15,000 acres of wooded habitats, and nearly 102,000 acres of grasslands or farmlands. Tourist accommodations exist at Nebraska City and Syracuse.

A. Federal Areas: None

B. State Areas

1. Wilson Creek WMA (Map location 1). Area 41 acres. Located one mile south and three miles east of Otoe. Includes a 14-acre reservoir and surrounding grasses and shrubs.

2. Arbor Lodge State Historical Park (Map location 2). Located just west of Nebraska City. There are hundreds of planted but mature trees in the arboretum, and a 0.5-mile "tree trail" with identification tags. Worth visiting for historical reasons as well as for learning tree identification. State park entry permit required.
http://outdoornebraska.ne.gov/parks/guides/park-search/showpark.asp?Area_No=4

3. Missouri River Basin Lewis and Clark Interpretive Trail and Visitor Center (Map location 3). This new center is located on 80 acres overlooking the Missouri River, southeast edge of Nebraska City (off U.S. Highway 2). It documents the natural history aspects of the 1804-1806 Expedition of Discovery.
http://mrb-lewisandclarkcenter.org/

4. Triple Creek WMA (Map location 4). Area 80 acres. Located three miles south and one mile west of Palmyra. Contains two intermittent streams and 16 acres of wooded habitats.

20. Saline County

Saline County is in a region of loess and dissected plains, with about 1,900 acres of surface water, almost 11,000 acres of wooded habitats, and about 89,000 acres of grasslands or farmlands. There are tourist accommodations at Crete, Friend and Wilber.

A. Federal Areas: None

B. State Areas

1. Swan Creek WMA. Area 160 acres. Located nine miles south and one mile east of Friend. Consists of a 27-acre lake, marshland, native prairie, wooded habitats and cropland.

2. Walnut Creek Public Use Area. 64 acres. Located three miles north and two miles east of Crete.

Various upland habitats are present.

21. Jefferson County

Jefferson County is in a region of loess and dissected plains, with nearly 2,000 acres of surface water, over 10,000 acres of wooded habitats, and 148,000 acres of grasslands or farmlands. There are tourist accommodations at Fairbury.

A. Federal Areas: None

B. State Areas

1. Alexandria Lakes SRA/WMA. Area 778 acres. Located nine miles west and six miles north of Fairbury, or 4 miles east of Alexandria. Consists of a 43-acre lake, marshes, streams, ponds and wooded habitats.

2. Rock Creek Station State Historical Park & SRA. Area 393 acres. Located about two miles north and three miles east of Endicott; see state highway map for location. State park entry permit required. The area includes about 100 acres of native prairie.

3. Rock Glen WMA. Located seven miles east and two miles south of Fairbury. Includes 706 acres of rolling native upland prairie and tree-lined drainages.

22. Gage County (Map 40)

Gage County is in an area of loess and glacial drift, with nearly 2,000 acres of surface water, 17,000 acres of wooded habitats, and 166,000 acres of grasslands or farmlands. Tourist accommodations are at Beatrice and Wymore.

A. Federal Areas

1. Homestead National Monument of America (Map location 2). This monument celebrates the Homestead Act. It includes a 2.5 mile trail passing through riparian wooded habitats and restored prairie. There is a bird checklist of more than 150 species, a plant list, and an educational exhibit that features pioneer history and related artifacts. A stand of restored native prairie is present. Phone 402/223-3514 for information.
www.nps.gov/home/index.htm

B. State Areas

1. Clatonia Public Use Area (Map location 1). Area 115 acres, with a 40-acre reservoir.

2. Iron Horse Trail (Map location 3). Several units, varied areas. There are more than 20 such sites in Gage and Pawnee counties, ranging from 1-19 acres.

3. Rockford Lake SRA (Map location 4). Area 436 acres, with a 150-acre reservoir. State park entry permit required.
http://outdoornebraska.ne.gov/parks/guides/park-search/showpark.asp?Area_No=154

4. Wolf Wildcat Public Use Area (Map location 5). Area 160 acres, with a 42-acre reservoir.

5. Arrowhead WMA (Map location 6). Area 320 acres. Upland habitats.

6. Diamond Lake WMA (Map location 7). Area 320 acres, including lakes. Diamond Lake consists of open grasslands and hardwood stands around a 33-acre reservoir. The adjoining Donald Whitney Memorial WMA is much smaller, and also includes a reservoir.

7. Big Indian Public Use Area. (Map location 8). Area 300 acres, including a 77-acre lake.

23. Johnson County (Map 41)

Johnson County is in a region of glacial drift, with about 500 acres of surface water, 6,800 acres of wooded habitats, and over 92,000 acres of grassland. Tourist accommodations are at Tecumseh.

A. Federal Areas: None

B. State Areas

1. Osage WMA (Map locations 1-3). Area 778 acres. Mostly comprised of wooded habitats and intervening grassland habitats, plus tree plantings and crops.

2. Hickory Ridge WMA (Map location 4). Area 250 acres. Includes 60 acres of timber, a small pond, and fairly steep grassland and creek-bottom habitat.

3. Twin Oaks WMA. (Map locations 5-7). Area 795 acres. Mostly wooded habitats, with some grassland and food plots.

24. Nemaha County (See Map 43)

Nemaha County is a Missouri Valley county with nearly 1,800 acres of surface water, over 21,000 acres of wooded habitats, and 48,000 acres of grasslands or farmlands. There are tourist accommodations at Auburn and Brownville.

A. Federal Areas: None

B. State Areas
1. Indian Cave State Park (Map location 1). See Richardson County.
http://outdoornebraska.ne.gov/parks/guides/park-search/showpark.asp?Area_No=91

2. Brownville SRA. Area 220 acres. Located at southeastern edge of Brownville, and provides boating access to the Missouri River. State park entry permit required.
http://outdoornebraska.ne.gov/parks/guides/park-search/showpark.asp?Area_No=36

25. Pawnee County (Map 42)

Pawnee County is in a region of glacial till, with less than 1,000 acres of surface water, over 16,000 acres of wooded habitats, and about 124,000 acres of grasslands, the latter floristically associated with the Flint Hills prairies of Kansas. There are tourist accommodations at Pawnee City and Steinauer.

A. Federal Areas: None

B. State Areas
1. Iron Horse Trail (Map location 1). Many short sections of railroad bed that are being used as a hike-bike trail. See also Gage County.

2. Burchard Lake WMA (Map location 2, plus enlarged inset). Area 560 acres, with a 150-acre reservoir, surrounded by native grasslands and some hardwoods. There is a small resident flock of greater prairie-chickens, and two permanent blinds are located on the hilltop that is used as a lek by males (see inset arrowhead).

3. Bowwood WMA (Map location 3). Area 320 acres, and comprised of wooded areas, croplands, grasslands and two small ponds.

4. Pawnee Prairie WMA (Map location 4). Area 1,021 acres. Comprised mostly of native prairie, with some wooded habitats and a small amount of cropland. Supports a flock of about 20 greater prairie-chickens, which have a lek located near the center of the prairie (about 0.75 miles from the various parking lots, which are at the east, northwest and southwest corners of the area). No permanent blinds are present, but temporary blinds are permitted. Driving on the prairie is not allowed.

5. Prairie Knoll WMA. (Map location 5). Area 120 acres. Includes a small reservoir and a mixture of wooded habitats, tree plantings, grassland and cropland.

6. Mayberry WMA (Map location 6). Located five miles north of Burchard and 0.5 miles east of highway 4. Consists of 195 acres of grassland, trees and a small reservoir.

7. Table Rock WMA. Located just east of Table Rock, on the north side of US Highway 4. Includes mainly wooded bottomland along the Nemaha River, plus grassland and cropland.

26. Richardson County (Map 43)

Richardson County is a Missouri Valley county with over 1,600 acres of surface water, nearly 16,000 acres of wooded habitats, and 110,000 acres of grassland. There are tourist accommodations at Falls City.

A. Federal Areas: None

B. State Areas
1. Indian Cave State Park. (Map location 1). Area 3,052 acres. Like Rulo Bluffs (see below), this area has a diverse wooded habitats flora of southern affinities, and supports such species as summer tanagers, Acadian flycatchers and chuck-will's-widows. The park is 78 percent mature forest, and the rest grassland or developed areas, with both modern and primitive camping facilities and about 20 miles of hiking trails, including a 14-mile "Hard-

wood Trail" that is moderately difficult. Classified as a Nebraska Important Bird Area. Park entry permit required. Phone 402/883-2575 for information. http://outdoornebraska.ne.gov/parks/guides/park-search/showpark.asp?Area_No=91

2. Verdon Lake SRA (Map location 2). Area 30 acres, including a small reservoir. State park entry permit required .

3. Kinters Ford WMA (Map location 3). Area 200 acres. Includes riverbottom wooded habitats, grassland and cropland.

4. Four Mile Creek WMA (Map location 4). Area 160 acres. Mostly upland habitats along a creek-bottom.

5. Iron Horse Trail (Map location 5). Area 210 acres. This hiking trail follows an old railroad right-of-way. See also Gage and Pawnee counties.

6. Margrave WMA (Map location 6). Located three miles south and seven miles east of Falls City. Consists of 106 acres along the Nemaha River, including wooded habitats, cropland, grasses and marshy areas.

C. Other Areas
1. Rulo Bluffs Preserve (Map location 7). Area 424 acres. This area, owned by the Nature Conservancy, is located nearly on the Kansas border, and has the most southern floral affinities of any Nebraska forest. Ridgetop prairie-savanna. There is no bird checklist yet available, but there should be some southern species in addition to chuck-will's-widow to be found there. Permission to visit must be obtained from the TNC Field Office in Omaha (402/342-0282).
www.nature.org/wherewework/northamerica/states/nebraska/preserves/art280.html

Tall-grass Prairie Sites in Eastern Nebraska

Birders who wish to find tall grass prairie birds in eastern Nebraska should consider these sites:

Boone County (p. 42)
Olsen Nature Preserve. 120 acres of tallgrass prairie, plus riparian cottonwoods and oaks along Beaver Creek. Located 8 miles north of Albion on Highway 14, than go left on gravel road for 1 mile. Owned by Prairie Plains Resource Inst., Aurora.

Cedar County (p. 57)
Wiseman WMA. Virgin prairie on ridges and hill-tops. See text description.

Dixon County (p. 57)
Buckskin Hills WMA. Small area of virgin prairie. See text description.
Ponca State Park. Small stands of virgin prairie on ridges and hilltops. See text description.

Douglas County (p. 61)
Stolley Prairie. 0.25 acres of virgin prairie in middle of Omaha's Northwest Park, east side of 168th St., between Dodge and Blondo.
Zorinsky Lake. 40 acres of virgin prairie; walking trail leads to prairie. See text description.

Gage County (p. 66)
Homestead National Monument. 100 acres of restored prairie. See text description.
Wildcat Creek Prairie. 36 acres upland prairie. From Virginia, 7 miles south, 1 mile west, 1 mile north, ¼ mile west. 402/488-9032

Jefferson County (p. 66)
Rock Creek Station State Historical Park & Rock Glen WMA. Park has over 100 acres of virgin and restored prairie, Rock Glen has several hundred acres of virgin and restored prairie. See text description.
Rose Creek WMA. About 200 acres of oak savanna, 8 miles southwest of Fairbury. (402/749-7650).

Lancaster County (p. 63)
Branched Oak WMA. Includes 200 acres of virgin prairie. See text description.
Highway 77 (Hike) Prairie. 20 acres of virgin prai-

rie, 8 miles south of Lincoln, northeast corner of Hwy 77 and Hickman Rd.

Little Salt Fork Marsh Preserve and Little Salt Creek WMA. Total of 336 acres of saline wetlands and meadows. Corner of 1st & Raymond Road (Preserve is on north side of road, WMA ¼ mile south).

Nine Mile Prairie. 230 acres virgin tallgrass prairie. See text description.

Pioneers Park Nature Center. 500 acres of restored and virgin prairie. See text description.

Wilderness Park. Includes small area of prairie on sandstone, west edge of park, 0.2 miles south of Pioneers Blvd. See text description.

Audubon Spring Creek Prairie. See text description.

Otoe County (p. 65)
Dieken Prairie. 14 acres of virgin prairie, From west edge of Unadilla 1 ¼ miles south, ¾ mile west. (402/488-9032)

Pawnee County (p. 67)
Burchard Lake WMA. Circa 350 acres of grassland, including prairie. See text description.

Klapka Farm. Includes circa 35 acres of prairie on 400-acre farm. 2 miles south of Table Rock (402/471-9032)

Pawnee Prairie WMA. Some virgin & restored prairie. See text description.

Richardson County (p. 67)
Indian Cave State Park. Ca. 40 acres of virgin prairie on hilltops and hay meadows. SE part of park, and along Trail 10 from top of bluffs to Adirondack shelter. See text description.

Saunders County (p. 61)
Jack Sinn WMA. Saline wetlands and lowland grasslands. See text description.

Red Cedar Recreation Area. Native and re-seeded grasslands. See text description.

Storm Prairie. 30 acres of lowland meadows. From Hwy 92 on east side of Yutan go 2 miles north, 1 mile east, ½ mile north. 402/488-9032.

Seward County (p. 62)
Bur Oak WMA. Ca. 40 ac. of prairie in oak woodland. See text description.

Twin Lakes WMA. Some virgin & restored prairie; best prairie SW of small lake, on west side of WMA. See text description.

Stanton County (p. 58)
Wood Duck WMA. Virgin prairie and prairie restoration. See text description.

Washington County (p. 60)
Boyer Chute National Wildlife Refuge. Ca. 2,000 acres of re-seeded prairie. See text description.

Cuming City Cemetery & Nature Preserve. 11 acres of virgin prairie. From intersection of Hwys 30 & 75 in Blair go 3.5 miles north on Hwy 75, turn left on County Rd. 14, then 2 blocks to cemetery.

DeSoto National Wildlife Refuge. Ca. 1900 acres of restored prairie. See text description.

Wayne County (p. 58)
Thompson-Barnes WMA. 18 acres of restored prairie, From Wayne go 3.5 miles north on Hwy 15, 1 mile west.

Greater Prairie-chicken

References

American Ornithologists' Union (AOU). 1998. *The A.O.U. Checklist of North American Birds.* 7th ed. AOU, Washington, D.C. Supplements in *Auk* 117:847-858; 119:697-906; 120:923-931.

Boyle, W. J., & R. H. Bauer. 1994. Birdfinding in Forty National Forests and Grasslands. *Birding* (supplement) 26(2): 1-186. (Includes Oglala National Grassland.)

Bray, T. E., B. K. Padelford and W. R. Silcock. 1986. *The Birds of Nebraska: a critically evaluated list.* Bellevue: Published by the authors. 109 pp.

Brogie, M. A. 1997-2005. (Reports of the NOU Records Committee.) *Nebraska Bird Review.* 65:115-126; 66:147-159; 67:141-152; 71:136-142; 72:59-65; 73:78-84.

——— & M. J. Mossman. 1983. Spring and summer birds of the Niobrara Valley Preserve area, Nebraska. *Nebraska Bird Review* 51:44-51.

Brown, C. R., M. B. Brown, P. A. Johnsgard, J. Kren & W. C. Scharf. 1996. Birds of the Cedar Point Biological Station area, Keith and Garden Counties, Nebraska: Seasonal occurrence and breeding data. *Transactions of the Nebraska Academy of Sciences* 29:91-108.

———, and M. B. Brown. 2001. Birds of the Cedar Point Biological Station. Occasional Papers of the Cedar Point Biological Station No.1. 36 pp.

Bruner, L. R., H. Wolcott & M. H. Swenk. 1904. *A preliminary review of the birds of Nebraska,* Omaha.

Busby, W. H. and J. L. Zimmerman. 2001. *Kansas Breeding Bird Atlas.* University Press of Kansas, Lawrence, KS.

Canterbury, J., and P. A. Johnsgard. 2000. A century of breeding birds in Nebraska. *Nebraska Bird Review* 68:89-101.

Colt, C. J. 1996. Breeding bird use of riparian forests along the central Platte River: A spatial analysis. M.S. thesis, University of Nebraska-Lincoln, Lincoln, NE. 104 pp.

Currier, P. J., G. R. Lingle, and J. G. VanDerwalker. 1985. *Migratory Bird Habitat on the Platte and North Platte Rivers in Nebraska.* Whooping Crane Habitat Maintenance Trust, Grand Island, NE. 177 pp.

Davis, C. A. 2005a. Breeding bird communities in riparian forests along the central Platte River, Nebraska. *Great Plains Research* 15:199-211.

———. 2005b. Breeding and migration bird use of a riparian woodland along the Platte River in central Nebraska. *North American Bird Bander,* July-September, 2005, pp. 109-114.

Ducey, J. E. 1988. *Nebraska Birds: Breeding Status and Distribution.* Omaha: Simmons-Boardman Books.

———. 2000. *Birds of the Untamed West. The History of Birdlife in Nebraska 1750-1875.* Making History Press. Omaha, NE.

Faanes, C. E., and G. R. Lingle. 1995. Breeding birds of the Platte Valley of Nebraska. Northern Prairie Wildlife Research Center Home Page, Jamestown, ND. URL: www.npwrc.usgs.gov/resource/birds/platte/index.htm (January 12, 2011)

Farrar, J. (ed.) 1985. Birds of Nebraska. *NEBRASKAland* (special issue) 63(1). 146 pp.

———. 2004. Birding Nebraska. *NEBRASKAland* (special issue) 82(1), 178 pp.

Fiala, K. L. 1970. The birds of Gage County, Nebraska. *Nebraska Bird Review* 38:42-72.

Haecker, F. W., R. A. Moser and J. B. Swenk. 1945. Check-list of the birds of Nebraska. *Nebraska Bird Review* 13:1-40.

Jacobs, B. 2001. *Birds in Missouri.* Missouri Dept. of Conservation, Jefferson City, MO.

Johnsgard, P. A. 1979. *Birds of the Great Plains: Breeding Species and their Distribution.* Lincoln: University of Nebraska Press.

———. 1980. *A Preliminary List of the Birds of Nebraska and Adjacent Plains States.* Occasional Papers (No.6) of the Nebraska Ornithologists' Union, Lincoln, Nebraska.

———. 1984. *The Platte: Channels in Time.* Lincoln, University of Nebraska Press.

———. 1996. *This Fragile Land: A Natural History of the*

Nebraska Sandhills. Lincoln: University of Nebraska Press.

——. 1998. Endemicity and regional biodiversity in Nebraska's breeding avifauna. *Nebraska Bird Review* 66:115-120.

——. 2000. Historic birds of Lincoln's Salt Basin and Nine-mile Prairie. *Nebraska Bird Review*. 68:132-136.

——. 2001. A century of ornithology in Nebraska: A personal view. Pp. 329-55, in *Contributions to the History of North American Ornithology*, Vol. II. (W. E. Davis & J. A. Jackson, eds.) Nuttall Ornithological Club, Boston, Mass.

——. 2001. *Prairie Birds: Fragile Splendor in the Great Plains*. University Press of Kansas, Lawrence.

——. 2001. *The Nature of Nebraska: Ecology and Biodiversity*. University of Nebraska Press, Lincoln.

——. 2001. Ecogeographic aspects of greater prairie-chicken leks in southeastern Nebraska. *Nebraska Bird Review*. 68:179-184.

——. 2002. The falconiform and strigiform fauna of Nebraska. *Nebraska Bird Review* 69:80-84.

——. 2003. Nebraska's sandhill crane populations, past, present and future. *Nebraska Bird Review* 70:175-177.

——. 2005. Habitat associations of Nebraska birds. *Nebraska Bird Review*, 73:20-25. (with John Dinan).

Johnson, W. C. 1961. Woodland expansion in the Platte River, Nebraska: Patterns and causes. *Ecological Monographs* 64:45-84.

Jorgensen. J. 2001-2003 (Reports of the NOU Records Committee.) *Nebraska Bird Review*. 69:85-91, 70:84-90, 71:97-201

——. 2004. *An Overview of Shorebird Migration in the Eastern Rainwater Basin*. Occasional Paper No.8, Nebraska Ornithologists' Union, Lincoln, Ne. 68 pp.

Kingery, H. (ed.). 1998. *Colorado Breeding Bird Atlas*. Colorado Division of Wildlife, Denver. 600 pp.

Knue, J. 1997. *Nebraskaland Magazine Wildlife Viewing Guide*. NEBRASKAland 75(1). 96 pp.

Krapu, G. (ed.). 1981. *The Platte River Ecology Study:* Special Research Report, Northern Prairie Wildlife Research Station, U.S. Fish & Wildlife Service, Jamestown, ND. 186 pp.

Lingle, G. R. 1994. *Birding Crane River: Nebraska's Platte*. Grand Island: Harrier Publ. Co.

Mollhoff, W. J. 2001. *The Nebraska Breeding Bird Atlas*. Nebraska Game & Parks Commission, Lincoln.

Nebraska Game & Parks Commission. 1972. *The Nebraska Fish and Wildlife Plan. Vol. 1. Nebraska Wildlife Resources Inventory*. Lincoln: Nebraska Game & Parks Commission. 242 pp. (This resource provides estimates of habitat acreages for individual counties and larger regions).

Nebraska Ornithologists' Union Records Committee. 1997. The official list of the birds of Nebraska. *Nebraska Bird Review* 65:3-16. (See Brogie & Jorgensen references for annual supplements.)

Peterson, R. A. 1995. *The South Dakota Breeding Bird Atlas*. South Dakota Ornithologists' Union, Aberdeen, SD.

Pettingill, O. S. Jr. 1981. *A Guide to Bird-finding West of the Mississippi*. 2nd. ed. New York: Oxford University Press.

Price, J., S. Droege, & A. Price. 1995. *The Summer Atlas of North American Birds*. San Diego: Academic Press.

Rapp, W. F. Jr. , J. L. C. Rapp, H. E. Baumgartner, & R. A. Moser. 1958. Revised checklist of Nebraska birds. *Occasional Papers, Nebraska Ornithologists' Union*, No.5, Lincoln, Nebraska. 36 pp.

Rosche, R. C. 1982. *Birds of Northwestern Nebraska and Southwestern South Dakota*. Chadron, Nebraska: Published by the author. 100 pp.

——. 1990. Birding pristine Nebraska. *Winging It* 2(6): 1-6.

——. 1994a. *Birds of the Lake McConaughy area and the North Platte Valley, Nebraska*. Chadron, Nebraska: Published by the author. 115 pp.

——. 1994b. Birding in western Nebraska. *Birding*, 26:179-189; 416-423.

——. & P. A. Johnsgard. 1984. Birds of Lake McConaughy and the North Platte Valley, Oshkosh to Keystone. *Nebraska Bird Review* 52: 26-35.

Sharpe, R. S., W. R. Silcock, and J. G. Jorgensen. 2001. *Birds of Nebraska: Their Distribution and Temporal Occurrence*. University of Nebraska Press, Lincoln.

Steinauer, G., and S. Rolfsmeier, 2003. *Terrestrial Natural Communities of Nebraska (Version III)*. Nebraska Game & Parks Commission, Lincoln, NE. 162 pp.

Tallman, D. A., D. L. Swanson & J. S. Palmer. 2002. *The Birds of South Dakota*. 3rd ed. South Dakota Ornithologists' Union, Aberdeen, SD.

Thompson, M. C. & C. Ely. 1989, 1992. *Birds in Kansas*. 2 vols. University of Kansas Press, Lawrence.

U.S. Fish & Wildlife Service. 1981. The Platte River ecology study: Special Research Report. Jamestown: Northern Prairie Wildlife Research Station. 186 pp.

County Index

Whooping Crane

Annotated Checklist of Regularly Occurring Nebraska Birds

This species list of 366 species excludes accidental, extinct, extirpated and hypothetical bird species. The family sequence and species names follow the American Ornithologists' Union (2004).

Inclusive arrival and departure dates for migrants are based on median dates, not extreme records. Early, mid- and late month designations represent 10-day units for better-documented migrants; migration periods for rare species are specified as to nearest month only. Migrant abundance status is based on published migration dates (extending from 1933 to 1980) tallied in Paul Johnsgard's *Birds of Nebraska* as follows: very common = 100+ total records, uncommon = 76-99 total records, occasional = 51-75 total records, rare = 25-50 total records, very rare = under 25 records. Clearly out-of-range species having very few state records are termed "vagrants."

Asterisks indicate known breeding species. Breeding status is based on total numbers of confirmed state records for each species as listed in *The Nebraska Breeding Bird Atlas* (2001): abundant = 150+, common = 101-149, uncommon = 75-100, scattered 51-75, local = 10-50, highly local = 1-9. "Historic" breeders are those that were not confirmed as breeding during the 1984 to 1989 atlasing period, but are known to have bred earlier and/or since then ("recent").

Family Anatidae – Swans, Geese and Ducks

Greater White-fronted Goose. *Anser albifrons.* Common spring (early March to mid-April) and fall (mid-October to mid-November) migrant.

Snow Goose. *Chen caerulescens.* Very common spring (early March to mid-April) and fall (early October to early December) migrant.

Ross's Goose. *Chen rossii.* Very rare (but increasing) spring (early March to mid-April) and fall (early October to early December) migrant.

Cackling (Canada) Goose. *Branta hutchinsii.* Com-mon spring (early March to mid-April) and fall (early October to early December) migrant. Of 404 Canada geese taken in Nebraska, 17 were classified as *hutchinsii* by Myron Swenk *(Nebraska Bird Review,* 2: 103-116; 1934); all were from Hall and Buffalo counties.

Canada Goose. *Branta canadensis.* Very common migrant and local permanent resident*

Trumpeter Swan. *Cygnus buccinator.* Extremely rare permanent resident (west).*

Tundra Swan. *Cygnus columbianus.* Rare migrant (mainly east), March and November.

Wood Duck. *Aix sponsa.* Very common migrant and uncommon breeding-season resident (mainly east), late March to late October.*

Gadwall. *Anas strepera.* Common migrant & local breeding-season resident, late March to November.*

Eurasian Wigeon. *Anas penelope.* Very rare migrant, mainly in spring.

American Wigeon. *Anas americana.* Very common migrant and highly local breeding-season resident, mid-March to late September.*

American Black Duck. *Anas rubripes.* Rare spring and fall migrant (east).

Mallard. *Anas platyrhynchos.* Very common migrant and scattered breeding-season resident, often overwintering.*

Blue-winged Teal. *Anas discors.* Very common migrant and local breeding-season resident, early April to mid-October.*

Cinnamon Teal. *Anas cyanoptera.* Uncommon migrant and highly local breeding-season resident (west), April to October.*

Northern Shoveler. *Anas clypeata.* Very common migrant and highly local breeding-season resident, March to November.*

Northern Pintail. *Anas acuta.* Very common migrant and local breeding-season resident, often overwintering.*

Green-winged Teal. *Anas crecca.* Very common migrant and highly local breeding-season resident, mid-March to early November.*

Canvasback. *Aythya valisineria.* Very common migrant and highly local breeding-season resident, mid-March to mid-November.*

Redhead. *Aythya americana.* Very common migrant and highly local breeding-season resident, mid-March to mid-November.*

Ring-necked Duck. *Aythya collaris.* Very common migrant and historic breeding-season resident (central), late-March to mid-October.*

Greater Scaup. *Aythya marila.* Rare overwintering migrant, late October to April.

Lesser Scaup. *Aythya affinis.* Very common migrant and historic breeding-season resident (central), mid-March to mid-October .*

Surf Scoter. *Melanitta perspicillata.* Very rare and fall migrant late April to late May, and October to December.

White-winged Scoter. *Melanitta fusca.* Rare spring and fall migrant, March to April, and October to December.

Black Scoter. *Melanitta nigra.* Very rare spring and fall migrant, March to May, and September to December.

Long-tailed Duck (Oldsquaw).*Clangula hyemalis.* Very rare spring and fall migrant, February to April and October to December.

Bufflehead. *Bucephala albeola.* Very common spring (mid-March to late April) and fall (mid-October to late November) migrant.

Common Goldeneye. *Bucephala clangula.* Very common spring (early March to early April) and fall (mid-November to mid-December) migrant.

Barrow's Goldeneye. *Bucephala islandica.* Very rare spring and fall migrant (west).

Hooded Merganser. *Lophodytes cucullatus.* Very common migrant, historic (and recent) breeding-season resident, late March to late November.*

Common Merganser. *Mergus merganser.* Very common spring (early March to late April) and fall (mid-November to mid-December) migrant, occasional overwintering and usually non-breeding (historic breeding) summer resident.*

Red-breasted Merganser. *Mergus serrator.* Very common spring (late March to late April) and fall (early to late November) migrant.

Ruddy Duck. *Oxyura jamaicensis.* Very common migrant and highly local breeding-season resident, early April to mid-November (central & west).*

Family Phasianidae – Partridges, Grouse and Turkeys

Gray Partridge. *Perdix perdix.* Historic (and recent) resident (northeast).*

Ring-necked Pheasant. *Phasianus colchicus.* Common resident.*

Greater Prairie-chicken. *Tympanuchus cupido.* Highly local resident (east and central).*

Lesser Prairie-chicken. *Tympanuchus pallidicinctus.* Extirpated.

Sharp-tailed Grouse. *Tympanuchus phasianellus.* Local resident (west and central).*

Wild Turkey. *Meleagris gallopavo.* Scattered (but increasing) resident.*

Family Odontophoridae – New World Quail

Northern Bobwhite. *Colinus virginianus.* Scattered resident.*

Family Gaviidae – Loons

Red-throated Loon. *Gavia stellata.* Very rare spring and fall migrant, April to May and October to December.

Pacific Loon. *Gavia pacifica.* Very rare spring and fall migrant, April to May and October to November.

Common Loon. *Gavia immer.* Very common spring (March to May) and fall (late October to early November) migrant, sometimes overwintering.

Family Podicipedidae – Grebes

Pied-billed Grebe. *Podilymbus podiceps.* Very common migrant and local breeding-season resident, early April to early November.*

Horned Grebe. *Podiceps auritus.* Very common spring (mid-April to early May) and fall (early October to mid-November) migrant, historic breeder.*

Red-necked Grebe. *Podiceps grisegena.* Very rare spring (April) and fall (October-November) migrant.

Eared Grebe. *Podiceps nigricollis.* Very common migrant and highly local breeding-season resident (west), mid-April to mid-October.*

Western Grebe. *Aechmophorus occidentalis.* Very common migrant and highly local breeding-season resident (central & west), early May to early October.*

Clark's Grebe. *Aechmophorus clarkii.* Very rare (but previously unrecognized) migrant, and historic (and recent) breeding-season resident (west), May to October.*

Family Pelecanidae – Pelicans

American White Pelican. *Pelecanus erythrorhynchos.* Very common migrant and non-breeding summer resident, late April to mid-October.

Family Phalacrocoracidae – Cormorants

Double-crested Cormorant. *Phalacrocorax auritus.* Very common migrant and highly local breeding-season resident, mid-April to late September.*

Neotropic Cormorant. *Phalacrocorax brasilianus.* Very rare spring and fall vagrant.

Family Ardeidae – Bitterns and Herons

American Bittern. *Botaurus lentiginosus.* Very common migrant and highly local breeding-season resident, early May to early October.*

Least Bittern. *Ixobrychus exilis.* Rare migrant and highly local breeding-season resident, mid-May to mid-August.*

Great Blue Heron. *Ardea herodias.* Very common migrant and local breeding-season resident, early April to mid-October.*

Great Egret. *Ardea alba.* Common migrant and non-breeding summer resident, historic breeder, late April to early September.*

Snowy Egret. *Egretta thula.* Uncommon migrant and irregular breeding-season resident, highly local breeder, early May to mid-August.*

Little Blue Heron. *Egretta caerulea.* Common migrant and irregular non-breeding summer resident, early May to mid-August.

Cattle Egret. *Bubulcus ibis.* Rare migrant and highly local breeding-season resident, rare breeder, early May to late August.*

Green Heron. *Butorides virescens.* Very common migrant and highly local breeding-season resident, late April to mid-September.*

Black-crowned Night-heron. *Nycticorax nycticorax.* Very common migrant and highly local breeding-season resident, late April to early September.*

Yellow-crowned Night-heron. *Nyctanassa violacea.* Uncommon migrant and historic breeding-season resident, early May to early September.*

Family Threskiornithidae - Ibises and Spoonbills

White-faced Ibis. *Plegadis chihi.* Rare migrant and highly local breeding-season resident (west), April to October.*

Family Cathartidae - American Vultures

Turkey Vulture. *Cathartes aura.* Very common migrant and highly local breeding-season resident, mid-April to late September.*

Family Accipitridae - Kites, Hawks, Eagles and Allies

Osprey. *Pandion haliaetus.* Very common spring (late April to early May) and fall (mid-September to mid-October) migrant.

Mississippi Kite. *Ictinia mississippiensis.* Very rare migrant, rare breeding-season resident (west), mid-May to mid-September.*

Bald Eagle. *Haliaetus leucocephalus.* Very common overwintering migrant, local breeding resident; mid-November to late March.*

Northern Harrier. *Circus cyaneus.* Uncommon migrant and local breeding-season resident, mid-March to early December.*

Sharp-shinned Hawk. *Accipiter striatus.* Very common overwintering migrant and possible highly local breeding-season resident, late November to late March.*

Cooper's Hawk. *Accipiter cooperii.* Very common wintering migrant and highly local breeding resident, mid-September to late April.*

Northern Goshawk. *Accipiter gentilis.* Uncommon wintering migrant, September to late April.

Red-shouldered Hawk. *Buteo lineatus.* Historic (and recent) permanent resident (southeast).*

Broad-winged Hawk. *Buteo platypterus.* Very common (but declining) spring (late April to mid-May) and fall (mid-September to early October) migrant, historic breeding-season resident (east).*

Swainson's Hawk. *Buteo swainsoni.* Very common migrant and local breeding-season resident (west), mid-April to late September.*

Red-tailed Hawk. *Buteo jamaicensis.* Local resident.*

Ferruginous Hawk. *Buteo regalis.* Highly local resident (west).*

Rough-legged Hawk. *Buteo lagopus.* Very common wintering migrant, early November to late March.

Golden Eagle. *Aquila chrysaetos.* Highly local resident (west).*

Family Falconidae - Falcons

American Kestrel. *Falco sparverius.* Scattered resident.*

Merlin. *Falco columbarius.* Very common overwintering migrant, highly local breeding-season resident (west), October to March.*

Prairie Falcon. *Falco mexicanus.* Highly local resident (west).*

Gyrfalcon. *Falco rusticolus.* Very rare wintering migrant (mainly central), November to March.

Peregrine Falcon. *Falco peregrinus.* Very common (but most records prior to 1960) overwintering migrant, historic (and recent) breeding-season resident (east), mid-September to late March.*

Family Rallidae - Rails, Gallinules and Coots

King Rail. *Rallus elegans.* Very rare migrant and historic breeding-season resident (east), May to August .*

Virginia Rail. *Rallus limicola.* Rare migrant and highly local breeding-season resident, early May to mid-September.*

Sora. *Porzana carolina.* Very common migrant and highly local breeding-season resident, early May to late September.*

Common Moorhen. *Gallinula chloropus.* Very rare migrant and highly local breeding-season resident (east), mid-May to late August.*

American Coot. *Fulica americana.* Very common migrant and local breeding-season resident, late March to early November.*

Family Gruidae - Cranes

Sandhill Crane. *Grus canadensis.* Very common spring (early March to mid-April) and fall (early October to early November) migrant; very rare breeding-season resident.*

Whooping Crane. *Grus americana.* Very rare spring (late March to early May) and fall (mid-September to early November) migrant (central).

Family Charadriidae - Plovers

Black-bellied Plover. *Pluvialis squatarola.* Very common spring (May) and fall (late August to early October) migrant.

American Golden-Plover. *Pluvialis dominica.* Common spring (May) and fall (late September to mid-October) migrant.

Snowy Plover. *Charadrius alexandrinus.* Very rare migrant and rare breeding-season resident, April to August.*

Semipalmated Plover. *Charadrius semipalmatus.* Very common spring (May) and fall (mid-August to mid-September) migrant.

Piping Plover. *Charadrius melodus.* Common migrant and local breeding-season resident, mid-May to mid-September.*

Killdeer. *Charadrius vociferus.* Very common migrant and abundant breeding-season resident, mid-March to mid-October*

Mountain Plover. *Charadrius montanus.* Very rare migrant and rare breeding-season resident (southwest), May to September.*

Family Recurvirostridae - Stilts and Avocets

Black-necked Stilt. *Himantopus mexicanus.* Very rare migrant and local breeding-season resident (west), April to August.*

American Avocet. *Recurvirostra americana.* Very common migrant and highly local breeding-season resident (central & west), late April to early September.*

Family Scolopacidae - Sandpipers and Phalaropes

Greater Yellowlegs. *Tringa melanoleuca.* Very common spring (mid-April to early May) and fall (mid-August to early October) migrant.

Lesser Yellowlegs. *Tringa flavipes.* Very common spring (mid-April to mid-May) and fall (mid-August to early October) migrant.

Solitary Sandpiper. *Tringa solitaria.* Very common spring (early to mid-May) and fall (early August to early September) migrant.

Willet. *Catoptrophorus semipalmatus.* Very common migrant and local breeding-season resident (central & west), late April to late August.*

Spotted Sandpiper. *Actitis macularia.* Very common migrant and local breeding-season resident, early May to early September.*

Upland Sandpiper. *Bartramia longicauda.* Very common migrant and scattered breeding-season resident, early May to late August.*

Whimbrel. *Numenius phaeopus.* Very rare spring (April to mid-May) migrant.

Long-billed Curlew. *Numenius americanus.* Very common migrant and local breeding-season resident (central & west), mid-April to mid-August.*

Hudsonian Godwit. *Limosa haemastica.* Uncommon spring (late April to mid-May) migrant (east).

Marbled Godwit. *Limosa fedoa.* Very common migrant and historic (and recent) breeding-season resident, late April to mid-September.*

Ruddy Turnstone. *Arenaria interpres.* Rare spring (April-May) migrant, very rare fall migrant (east).

Red Knot. *Calidris canutus.* Very rare spring (May) and fall (September) migrant (east).

Sanderling. *Calidris alba.* Common spring (early to mid-May) and fall (late August to early October) migrant.

Semipalmated Sandpiper. *Calidris pusilla.* Very common spring (late April to mid-May) and fall (early August to mid-September) migrant.

Western Sandpiper. *Calidris mauri.* Common spring (early to mid-May) and fall (mid-August to early September) migrant (mainly west).

Least Sandpiper. *Calidris minutilla.* Very common spring (May) and fall (early August to mid-September) migrant.

White-rumped Sandpiper. *Calidris fuscicollis.* Very common spring (late April to mid-May) and rare fall (August) migrant.

Baird's Sandpiper. *Calidris bairdii.* Very common spring (mid-April to mid-May) and fall (mid-August to early October) migrant.

Pectoral Sandpiper. *Calidris melanotos.* Common spring (late April to mid-May) and fall (September-October) migrant.

Dunlin. *Calidris alpina.* Occasional spring (May) and fall (September) migrant (east).

Stilt Sandpiper. *Calidris himantopus.* Very common spring (early to mid-May) and fall (mid-August to mid-September) migrant.

Buff-breasted Sandpiper. *Tryngites subruficollis.* Rare spring (May) and fall (September) migrant (east).

Short-billed Dowitcher. *Limnodromus griseus.* Rare spring (May) and fall (August-September) migrant (east).

Long-billed Dowitcher. *Limnodromus scolopaceus.* Common spring (early to mid-May) and fall (early August to mid-October) migrant.

Wilson's Snipe. *Gallinago delicata.* Very common migrant and highly local breeding-season resident, mid-April to mid-November.*

American Woodcock. *Scolopax minor.* Rare migrant and highly local breeding-season resident (east and central), mid-April to mid-October.*

Wilson's Phalarope. *Phalaropus tricolor.* Very common migrant and highly local breeding-season resident (west and central), early May to early September.*

Red-necked Phalarope. *Phalaropus lobatus.* Common spring (May) and fall (mid-August to late September) migrant.

Red Phalarope. *Phalaropus fulicaria.* Very rare fall migrant, August to October.

Family Laridae - Gulls and Terns

Laughing Gull. *Leucophaeus atricilla.* Very rare April to December vagrant.

Franklin's Gull. *Leucophaeus pipixcan.* Very common spring (mid-April to mid-May) and fall (early September to mid-October) migrant.

Bonaparte's Gull. *Chroicocephalus philadelphia.* Uncommon spring (April-May) and fall (September-October) migrant.

Ring-billed Gull. *Larus delawarensis.* Very common non-breeding resident.

California Gull. *Larus californicus.* Rare year round vagrant (west).

Herring Gull. *Larus argentatus.* Common non-breeding resident.

Thayer's Gull. *Larus thayeri.* Rare overwintering vagrant.

Iceland Gull. *Larus glaucoides.* Very rare overwintering vagrant.

Lesser Black-backed Gull. *Larus fuscus.* Very rare overwintering vagrant.

Glaucous Gull. *Larus hyperboreus.* Very rare overwintering vagrant (mainly west).

Great Black-backed Gull. *Larus marinus.* Very rare overwintering vagrant.

Caspian Tern. *Hydroprogne caspia.* Uncommon migrant (east); uncommon non-breeding summer resident (west), early May to mid-September.

Common Tern. *Sterna hirundo.* Common spring (May) and fall (September) migrant (east).

Forster's Tern. *Sterna forsteri.* Very common migrant and highly local breeding-season resident (west and central), late April to mid-September.*

Least Tern. *Sterna antillarum.* Common migrant and local breeding-season resident, late May to mid-August (east to west-central).*

Black Tern. *Chlidonias niger.* Very common migrant and highly local breeding-season resident (west and central), mid-May to early September.*

Family Columbidae - Pigeons and Doves

Rock Pigeon. *Columba livia.* Uncommon resident.*

Eurasian Collared-dove. *Streptopelia decaocto.* Self-introduced resident.*

Mourning Dove. *Zenaida macroura.* Abundant breeding-season resident, late March to early November.*

Family Cuculidae - Cuckoos and Anis

Black-billed Cuckoo. *Coccyzus erythropthalmus.* Very common migrant and very local breeding-season resident, late May to late August.*

Yellow-billed Cuckoo. *Coccyzus americanus.* Very common migrant and local breeding-season resident, late May to mid-September.*

Family Tytonidae - Barn Owls

Barn Owl. *Tyto alba.* Uncommon resident.*

Family Strigidae - Typical Owls

Eastern Screech-owl. *Megascops asio.* Local resident.*

Great Horned Owl. *Bubo virginianus.* Scattered resident.*

Snowy Owl. *Bubo scandiacus.* Rare wintering migrant, November to April.

Burrowing Owl. *Athene cunicularia.* Very common migrant and local breeding-season resident (west to central), late April to mid-September.*

Barred Owl. *Strix varia.* Highly local resident (east).*

Long-eared Owl. *Asio otus.* Highly local resident (mostly east).*

Short-eared Owl. *Asio flammeus.* Highly local resident.*

Northern Saw-whet Owl. *Aegolius acadicus.* Rare overwintering (November to February) migrant, possible rare breeding-season resident (northwest).

Family Caprimulgidae - Goatsuckers

Common Nighthawk. *Chordeiles minor.* Very common migrant and highly local breeding-season resident, late May to mid-September (east).*

Common Poorwill. *Phalaenoptilus nuttallii.* Uncommon migrant and highly local breeding-season resident, early May to early September (west).*

Chuck-will's-widow. *Caprimulgus carolinensis.* Very rare migrant and highly local breeding-season resident, May to August (east, southeast).*

Whip-poor-will. *Caprimulgus vociferus.* Uncommon migrant and highly local breeding-season resident, early May to early September (east).*

Family Apodidae - Swifts

Chimney Swift. *Chaetura pelagica.* Very common migrant and local breeding-season resident (mainly east), late April to early October.*

White-throated Swift. *Aeronautes saxatalis.* Rare migrant and highly local breeding-season resident (west), mid-May to late August.*

Family Trochilidae - Hummingbirds

Ruby-throated Hummingbird. *Archilochus colubris.* Very common migrant and highly local breeding-season resident (east), mid-May to mid-September.*

Calliope Hummingbird. *Stellula calliope.* Very rare, mostly fall (August, September) migrant (west).

Broad-tailed Hummingbird. *Selasphorus platycercus.* Very rare, mostly fall (August-September) migrant (west).

Rufous Hummingbird. *Selasphorus rufus.* Very rare fall (August) migrant (west).

Family Alcedinidae - Kingfishers

Belted Kingfisher. *Ceryle alcyon.* Local breeding-season resident, mid-March to mid-November.*

Family Picidae - Woodpeckers

Lewis's Woodpecker. *Melanerpes lewis.* Uncommon migrant and highly local breeding-season resident, May to September (west).*

Red-headed Woodpecker. *Melanerpes erythrocephalus.* Very common migrant and abundant breeding-season resident, May to mid-September.*

Red-bellied Woodpecker. *Melanerpes carolinus.* Local resident.*

Yellow-bellied Sapsucker. *Sphyrapicus varius.* Uncommon overwintering migrant, October to March (east).

Downy Woodpecker. *Picoides pubescens.* Scattered resident.*

Hairy Woodpecker. *Picoides villosus.* Local resident.*

Northern Flicker. *Colaptes auratus.* Uncommon resident (yellow-shafted east, red-shafted west).*

Pileated Woodpecker. *Dryocopus pileatus.* Historic (and recent) local resident (southeast).*

Family Tyrannidae - Tyrant Flycatchers

Olive-sided Flycatcher. *Contopus cooperi.* Common spring (mid to late May) and fall (early to late September) migrant.

Western Wood-pewee. *Contopus sordidulus.* Uncommon migrant and highly local breeding-season resident, late May to early September (west).*

Eastern Wood-pewee. *Contopus virens.* Very common migrant and highly local breeding-season resident, mid-May to mid-September (east).*

Yellow-bellied Flycatcher. *Empidonax flaviventris.* Uncommon spring (May) and fall (September) migrant; hypothetical breeding resident (east).

Acadian Flycatcher. *Empidonax virescens.* Rare spring (May) and fall (August) migrant, historic breeding-season resident (southeast).*

Alder Flycatcher. *Empidonax alnorum.* Very rare spring (late May to early June) and fall (August) migrant.

Willow Flycatcher. *Empidonax traillii.* Common migrant and local breeding-season resident, mid-May to early September (east).*

Least Flycatcher. *Empidonax minimus.* Very common (early to mid-May) and fall (September) migrant; hypothetical breeding resident (north).*

Cordilleran Flycatcher. *Empidonax occidentalis.* Very rare migrant and highly local breeding-season resident (northwest), May to September.*

Eastern Phoebe. *Sayornis phoebe.* Very common migrant and common breeding-season resident, mid-April to late September.*

Say's Phoebe. *Sayornis saya.* Very common migrant and local breeding-season resident, mid-April to mid-September (west).*

Great Crested Flycatcher. *Myiarchus crinitus.* Very common migrant and local breeding-season resident, late April to early September.*

Cassin's Kingbird. *Tyrannus vociferans.* Very rare migrant and highly local breeding-season resident, early May to mid-September (west).*

Western Kingbird. *Tyrannus verticalis.* Very common migrant and common breeding-season resident, early May to early September.*

Eastern Kingbird. *Tyrannus tyrannus.* Very common migrant and abundant breeding-season resident, early May to early September.*

Scissor-tailed Flycatcher. *Tyrannus forficatus.* Rare migrant and highly local breeding-season resident, early May to mid-September (east).*

Family Laniidae – Shrikes

Loggerhead Shrike. *Lanius ludovicianus.* Very common migrant and scattered breeding-season resident, mid-April to mid-September.*

Northern Shrike. *Lanius excubitor.* Uncommon wintering migrant, early November to mid-March.

Family Vireonidae - Vireos

White-eyed Vireo. *Vireo griseus.* Uncommon migrant and historic breeding-season resident, mid-May to early September (southeast).*

Bell's Vireo. *Vireo bellii.* Very common migrant and local breeding-season resident, mid-May to early September.*

Plumbeous Vireo. *Vireo plumbeus.* Very rare migrant and highly local breeding-season resident (northwest), migration dates poorly documented, approximately mid-May to mid-September.*

Blue-headed Vireo. *Vireo solitarius.* Very common spring (early to mid-May) and fall (mid-September to early October) migrant (east).

Yellow-throated Vireo. *Vireo flavifrons.* Very common migrant and highly local breeding-season resident (east), early May to early September.*

Warbling Vireo. *Vireo gilvus.* Very common migrant

and local breeding-season resident, early May to early September.*

Philadelphia Vireo. *Vireo philadelphicus.* Common spring (mid to late May) and fall (late August to late September) migrant (east).

Red-eyed Vireo. *Vireo olivaceus.* Very common migrant and local breeding-season resident, mid-May to early September.*

Family Corvidae - Jays, Magpies and Crows

Blue Jay. *Cyanocitta cristata.* Uncommon resident.*

Pinyon Jay. *Gymnorhinus cyanocephalus.* Highly local resident (west).*

Clark's Nutcracker. *Nucifraga columbiana.* Highly local resident (west).*

Black-billed Magpie. *Pica pica.* Uncommon resident (west).*

American Crow. *Corvus brachyrhynchos.* Scattered resident.*

Family Alaudidae - Larks

Horned Lark. *Eremophila alpestris.* Common resident.*

Family Hirundinidae - Swallows

Purple Martin. *Progne subis.* Very common migrant and local breeding-season resident, mid-April to late August (mostly east).*

Tree Swallow. *Tachycineta bicolor.* Very common migrant and local breeding-season resident, late April to mid-September.*

Violet-green Swallow. *Tachycineta thalassina.* Rare migrant and highly local breeding-season resident, mid-May to late August (west).*

Northern Rough-winged Swallow. *Stelgidopteryx serripennis.* Very common migrant and uncommon breeding-season resident, late April to early September.*

Bank Swallow. *Riparia riparia.* Very common migrant and local breeding-season resident, early May to early September.*

Cliff Swallow. *Petrochelidon pyrrhonota.* Very common migrant and common breeding-season resident, late April to early September.*

Barn Swallow. *Hirundo rustica.* Very common migrant and abundant breeding-season resident, late April to late September.*

Family Paridae - Titmice

Black-capped Chickadee. *Poecile atricapillus.* Common resident.*

Tufted Titmouse. *Baeolophus bicolor.* Highly local resident (east).*

Family Sittidae - Nuthatches

Red-breasted Nuthatch. *Sitta canadensis.* Very common wintering migrant (early October to early April); highly local breeding-season resident (north).*

White-breasted Nuthatch. *Sitta carolinensis.* Local resident.*

Pygmy Nuthatch. *Sitta pygmaea.* Highly local resident (west).*

Family Certhiidae - Creepers

Brown Creeper. *Certhia americana.* Historic resident.*

Family Troglodytidae - Wrens

Rock Wren. *Salpinctes obsoletus.* Very common migrant and local breeding-season resident, early May to late October (west).*

Carolina Wren. *Thryothorus ludovicianus.* Highly local resident (east).*

Bewick's Wren. *Thryomanes bewickii.* Very common migrant and historic (and recent) breeding-season resident, late April to late September (southeast).*

House Wren. *Troglodytes aedon.* Very common migrant and common breeding-season resident, late April to late September.*

Winter Wren. *Troglodytes troglodytes.* Uncommon wintering migrant, mid-October to mid-April.

Sedge Wren. *Cistothorus platensis.* Rare migrant and highly local breeding-season resident, early May to late September (east).*

Marsh Wren. *Cistothorus palustris.* Very common migrant and local breeding-season resident, early May to early October.*

Family Regulidae - Kinglets

Golden-crowned Kinglet. *Regulus satrapa.* Very common wintering migrant, mid-October to mid-April.

Ruby-crowned Kinglet. *Regulus calendula.* Very common spring and fall migrant, April to mid-May and September to October, sometimes overwintering.

Family Sylviidae - Gnatcatchers

Blue-gray Gnatcatcher. *Polioptila caerulea.* Common migrant and highly local breeding-season resident, early May to early September (east).*

Family Turdidae - Thrushes and Allies

Eastern Bluebird. *Sialia sialis.* Very common migrant and scattered breeding-season resident, late March to early November.*

Mountain Bluebird. *Sialia currucoides.* Very common migrant and highly local breeding-season resident (west), recorded April to October.*

Townsend's Solitaire. *Myadestes townsendi.* Common wintering (September to March) migrant (west), highly local resident (northwest)*

Veery. *Catharus fuscescens.* Very common spring (May) and fall (September) migrant.

Gray-cheeked Thrush. *Catharus minimus.* Very common spring (May) and fall (September) migrant.

Swainson's Thrush. *Catharus ustulatus.* Very common spring (early to late May) and fall (September) migrant.

Hermit Thrush. *Catharus guttatus.* Very common spring (April to May) and fall (September) migrant.

Wood Thrush. *Hylocichla mustelina.* Very common migrant and highly local breeding-season resident, mid-May to mid-September (east, north).*

American Robin. *Turdus migratorius.* Very common migrant and abundant breeding-season resident, late February to mid-December, sometimes overwintering.*

Varied Thrush. *Ixoreus naevius.* Very rare spring and fall vagrant (west).

Family Mimidae - Mockingbirds and Thrashers

Gray Catbird. *Dumetella carolinensis.* Very common migrant and scattered breeding-season resident, mid-May to late September.*

Northern Mockingbird. *Mimus polyglottos.* Very common migrant and local breeding-season resident, early May to mid-September.*

Sage Thrasher. *Oreoscoptes montanus.* Very rare migrant and historic breeding-season resident, mid-April to mid-September (west).*

Brown Thrasher. *Toxostoma rufum.* Very common migrant and common breeding-season resident, late April to late September.*

Family Sturnidae - Starlings

European Starling. *Sturnus vulgaris.* Abundant resident.*

Family Motacillidae - Pipits

American Pipit. *Anthus rubescens.* Very common spring (April) and fall (October) migrant.

Sprague's Pipit. *Anthus spragueii.* Uncommon spring (April) and fall (September to October) migrant.

Family Bombycillidae - Waxwings

Bohemian Waxwing. *Bombycilla garrulus.* Rare wintering (late November to late February) migrant (north).

Cedar Waxwing. *Bombycilla cedrorum.* Very common migrant and highly local breeding-season resident, late February to early October, sometimes overwintering.*

Family Parulidae - Wood Warblers

Blue-winged Warbler. *Vermivora pinus.* Rare spring (May) and fall (August-September) migrant (east).

Golden-winged Warbler. *Vermivora chrysoptera.* Very rare spring (May) and fall (September) migrant (east).

Tennessee Warbler. *Oreothlypis peregrina.* Very common spring (early to late May) and fall (early September to early October) migrant.

Orange-crowned Warbler. *Oreothlypis celata.* Very common spring (late April to mid-May) and fall (mid-September to mid-October) migrant.

Nashville Warbler. *Oreothlypis ruficapilla.* Very common spring (early to mid-May) and fall (mid-September to early October) migrant.

Northern Parula. *Parula americana.* Rare migrant and historic (and recent) breeding-season resident, early May to mid-September (east).*

Yellow Warbler. *Dendroica petechia.* Very common

migrant and local breeding-season resident, early May to early September.*

Chestnut-sided Warbler. *Dendroica pensylvanica.* Common spring (mid to late May) and fall (early to late September) migrant (east).

Magnolia Warbler. *Dendroica magnolia.* Very common spring (mid to late May) and fall (early September to early October) migrant (east).

Cape May Warbler. *Dendroica tigrina.* Rare spring (May) and fall (September) migrant (east).

Black-throated Blue Warbler. *Dendroica caerulescens.* Rare spring (May) and fall (September) migrant (east).

Yellow-rumped Warbler. *Dendroica coronata.* Very common migrant and highly local breeding-season resident, late April to late October (northwest).*

Black-throated Green Warbler. *Dendroica virens.* Common spring (early to mid-May) and fall (mid-September to early October) migrant (east).

Townsend's Warbler. *Dendroica townsendi.* Very rare spring (May) and fall (September) migrant (west).

Blackburnian Warbler. *Dendroica fusca.* Very common spring (May) and fall (early September to early October) migrant (east).

Yellow-throated Warbler. *Dendroica dominica.* Rare migrant and historic (and recent) breeding-season resident, early May to early September (southeast).*

Pine Warbler. *Dendroica pinus.* Very rare spring (May) and fall (September) migrant (east).

Palm Warbler. *Dendroica palmarum.* Uncommon spring (May) and fall (September) migrant (east).

Bay-breasted Warbler. *Dendroica castanea.* Common spring (May) and fall (September) migrant (east).

Blackpoll Warbler. *Dendroica striata.* Very common spring (May) and fall (September) migrant.

Cerulean Warbler. *Dendroica cerulea.* Rare migrant, historic (and recent) breeding-season resident, mid-May to mid-August (east).*

Black-and-white Warbler. *Mniotilta varia.* Very common migrant, and historic (and recent) breeding-season resident, early May to mid-September (north).*

American Redstart. *Setophaga ruticilla.* Very common migrant and highly local breeding-season resident, mid-May to mid-September (north).*

Prothonotary Warbler. *Protonotaria citrea.* Rare migrant and highly local breeding-season resident, mid-May to mid-September (southeast).*

Worm-eating Warbler. *Helmitheros vermivorus.* Very rare spring (May) and fall (September) migrant (east).

Ovenbird. *Seiurus aurocapillus.* Very common migrant and highly local breeding-season resident, mid-May to mid-September (east & north).*

Northern Waterthrush. *Parkesia noveboracensis.* Very common spring (May) and fall (late August to late September) migrant (mainly east).

Louisiana Waterthrush. *Parkesia motacilla.* Common migrant, highly local breeding-season resident, early May to late August (east).*

Kentucky Warbler. *Oporornis formosus.* Uncommon migrant and highly local breeding-season resident, mid-May to late August (southeast).*

Connecticut Warbler. *Oporornis agilis.* Rare spring (May) and fall (September) migrant (east).

Mourning Warbler. *Oporornis philadelphia.* Very common spring (mid to late May) and fall (early September to early October) migrant (east).

MacGillivray's Warbler. *Oporornis tolmiei.* Rare spring (May) and fall (September) migrant (west).

Common Yellowthroat. *Geothlypis trichas.* Very common migrant and local breeding-season resident, early May to mid-September.*

Hooded Warbler. *Wilsonia citrina.* Rare spring (May) and fall (August) migrant (east).

Wilson's Warbler. *Wilsonia pusilla.* Very common spring (May) and fall (September) migrant (mainly east).

Canada Warbler. *Wilsonia canadensis.* Uncommon spring (May) and fall (September) migrant (east).

Yellow-breasted Chat. *Icteria virens.* Very common migrant and highly local breeding-season resident, mid-May to early September.*

Family Thraupidae - Tanagers

Summer Tanager. *Piranga rubra.* Rare migrant and highly local breeding-season resident, mid-May to September (southeast).*

Scarlet Tanager. *Piranga olivacea.* Very common migrant and highly local breeding-season resident, mid-May to late August (east).*

Western Tanager. *Piranga ludoviciana.* Common migrant and highly local breeding-season resident (west), mid-May to mid-September.*

Family Emberizidae - Towhees, Sparrows and Longspurs

Green-tailed Towhee. *Pipilo chlorurus.* Rare spring (May) and fall (September) migrant (west).

Eastern Towhee. *Pipilo erythrophthalmus.* Very common migrant and local breeding-season resident, late April to mid-October (east and central).*

Spotted Towhee. *Pipilo maculatus.* Very common migrant and local breeding-season resident, late April to mid-October (west and central).*

Cassin's Sparrow. *Peucaea cassinii.* Very rare migrant, and historic (and recent) breeding-season resident (west), migration period undocumented.*

American Tree Sparrow. *Spizella arborea.* Very common wintering migrant, late October to early April.

Chipping Sparrow. *Spizella passerina.* Very common migrant and local breeding-season resident, late April to early October.*

Clay-colored Sparrow. *Spizella pallida.* Very common spring (early to mid-May) and fall (mid-September to early October) migrant.

Brewer's Sparrow. *Spizella breweri.* Rare migrant, and historic (and recent) breeding-season resident, early May to early September (northwest).*

Field Sparrow. *Spizella pusilla.* Very common migrant and local breeding-season resident, mid-April to early October.*

Vesper Sparrow. *Pooecetes gramineus.* Very common migrant and highly local breeding-season resident, mid-April to early October.*

Lark Sparrow. *Chondestes grammacus.* Very common migrant and uncommon breeding-season resident, early May to early September.*

Lark Bunting. *Calamospiza melanocorys.* Very common migrant and local breeding-season resident, early May to late August (west).*

Savannah Sparrow. *Passerculus sandwichensis.* Very common migrant and historic (and recent) breeding-season resident, late April to mid-September (northwest).*

Grasshopper Sparrow. *Ammodramus savannarum.* Very common migrant and uncommon breeding-season resident, early May to early September.*

Baird's Sparrow. *Ammodramus bairdii.* Uncommon spring (April-May) and fall (late September to mid-October) migrant (mainly west).

Henslow's Sparrow. *Ammodramus henslowii.* Rare migrant and historic (and recent) breeding-season resident, late April to late September (southeast).*

Le Conte's Sparrow. *Ammodramus leconteii.* Common spring (late April to early May) and fall (late September to late October) migrant.

Nelson's Sharp-tailed Sparrow. *Ammodramus nelsoni.* Very rare spring (May) and fall (October) migrant (east).

Fox Sparrow. *Passerella iliaca.* Very common spring (late March to mid-April) and fall (mid-October to mid-November) migrant.

Song Sparrow. *Melospiza melodia.* Very common migrant and local breeding-season resident, early April to late December.*

Lincoln's Sparrow. *Melospiza lincolnii.* Very common spring (late April to mid-May) and fall (mid-September to mid-October) migrant.

Swamp Sparrow. *Melospiza georgiana.* Uncommon migrant and highly local breeding-season resident, late April to late October (east and central).*

White-throated Sparrow. *Zonotrichia albicollis.* Very common wintering migrant, early October to mid-May.

Harris's Sparrow. *Zonotrichia querula.* Very common wintering migrant, mid-October to mid-May.

White-crowned Sparrow. *Zonotrichia leucophrys.* Very common wintering migrant, early October to mid-May.

Dark-eyed Junco. *Junco hyemalis.* Very common wintering migrant, early October to late March; also highly local breeding season resident (northwest).*

McCown's Longspur. *Rhyncophanes mccownii.* Rare migrant and highly local breeding-season resident, early April to early October (west).*

Lapland Longspur. *Calcarius lapponicus.* Common wintering migrant, mid-November to late February.

Chestnut-collared Longspur. *Calcarius ornatus.* Uncommon migrant, and historic (and recent) breeding-season resident, mid-April to early October (northwest).*

Snow Bunting. *Plectrophenax nivalis.* Rare wintering migrant, mid-November to mid-February.

Family Cardinalidae - Cardinals, Grosbeaks and Allies

Northern Cardinal. *Cardinalis cardinalis.* Local resident.*

Rose-breasted Grosbeak. *Pheucticus ludovicianus.* Very common migrant and local breeding-season resident, early May to mid-September (east and central).*

Black-headed Grosbeak. *Pheucticus melanocephalus.* Very common migrant and highly local breeding-season resident, mid-May to late August (west).*

Blue Grosbeak. *Passerina caerulea.* Very common migrant and local breeding-season resident, mid-May to late August.*

Lazuli Bunting. *Passerina amoena.* Very common migrant and highly local breeding-season resident, mid-May to late August (west and central).*

Indigo Bunting. *Passerina cyanea.* Very common migrant and local breeding-season resident, mid-May to late August (east and central).*

Dickcissel. *Spiza americana.* Very common migrant and scattered breeding-season resident, mid-May to late August (mainly east, local in west).*

Family Icteridae - Meadowlarks, Blackbirds, Orioles and Allies

Bobolink. *Dolichonyx oryzivorus.* Very common migrant and local breeding-season resident, mid-May to mid-August.*

Red-winged Blackbird. *Agelaius phoeniceus.* Very common migrant and abundant breeding-season resident, early March to late November.*

Eastern Meadowlark. *Sturnella magna.* Very common migrant and local breeding-season resident (mainly east), early April to mid-October.*

Western Meadowlark. *Sturnella neglecta.* Very common migrant and abundant breeding-season resident, early March to late October.*

Yellow-headed Blackbird. *Xanthocephalus xanthocephalus.* Very common migrant and scattered breeding-season resident, mid-April to mid-September.*

Rusty Blackbird. *Euphagus carolinus.* Common spring (late March to mid-April) and fall (early November to late December) migrant, frequently overwintering.

Brewer's Blackbird. *Euphagus cyanocephalus.* Common migrant and highly local breeding-season resident, mid-April to early November (west).*

Great-tailed Grackle. *Quiscalus mexicanus.* Rare (but increasing rapidly) migrant and highly local breeding-season resident, April to September (south).*

Common Grackle. *Quiscalus quiscula.* Very common migrant and abundant breeding-season resident, late March to late October.*

Brown-headed Cowbird. *Molothrus ater.* Very common migrant and uncommon breeding-season resident, mid-April to early October.*

Orchard Oriole. *Icterus spurius.* Very common migrant and uncommon breeding-season resident, mid-May to late August.*

Baltimore Oriole. *Icterus galbula.* Very common migrant and common breeding-season resident, early May to early September (east, hybridizing with Bullock's in central Nebraska).*

Bullock's Oriole. *Icterus bullockii.* Very common migrant and local breeding-season resident, early May to early September (west).*

Family Fringillidae - Finches

Gray-crowned Rosy-finch. *Leucosticte tephrocotis.* Very rare overwintering migrant, October to February (northwest).

Pine Grosbeak. *Pinicola enucleator.* Rare wintering migrant, late November to mid-March.

Purple Finch. *Carpodacus purpureus.* Common wintering migrant, late October to late April.

Cassin's Finch. *Carpodacus cassinii.* Very rare wintering migrant, October to mid-April.

House Finch. *Carpodacus mexicanus.* Local (now widespread) resident.*

Red Crossbill. *Loxia curvirostra.* Uncommon wintering migrant, mid-November to early April, highly local resident (northwest and west).*

White-winged Crossbill. *Loxia leucoptera.* Rare overwintering migrant, October to April.

Common Redpoll. *Acanthis flammeus.* Rare wintering migrant, late November to mid-March.

Pine Siskin. *Spinus pinus.* Common wintering migrant, mid-October to mid-May; irregular & highly local resident.*

American Goldfinch. *Carduelis tristis.* Local resident.*

Evening Grosbeak. *Coccothraustes vespertinus.* Common overwintering migrant, early November to late April (west).

Family Passeridae - Old World Sparrows

House Sparrow. *Passer domesticus.* Abundant resident.*

Calendar of Nebraska's Migratory Birds

These arrival dates of about 260 migratory species (those for which sample sizes were deemed adequate) are based on *Nebraska Bird Review* data originally summarized by Paul Johnsgard in *The Birds of Nebraska,* and encompass nearly five decades from the 1930s to 1980. Entire data summaries are available on-line at the Nebraska Ornithologists' Union (NOU) website: www.noubirds.org.

Median arrival and departure ("Depart.") dates were determined for the entire state, but most data came from observers in eastern Nebraska. Departure dates based on small sample sizes are indicated as *circa (ca.).* Spring arrivals in northwestern Nebraska tend to average about 7-10 days later than those in southeastern Nebraska, and fall departure dates average correspondingly earlier; the dates shown here probably best fit east-central Nebraska. Furthermore, recent warmer than normal years have increasingly influenced both arrivals and departures, resulting in somewhat earlier average spring migrations and later fall migrations than the dates shown here.

Median Spring Arrival/Departure Dates

February
20   American Robin (Depart. Nov. 19)
24   Cedar Waxwing (Depart. Oct. 4)

March
1   Sandhill Crane (Depart. Apr. 7)
3   Red-winged Blackbird (Depart. Nov. 21)
4   Western Meadowlark (Depart. Oct. 28)
9   Snow Goose (Depart. Apr. 20) Common Goldeneye (Depart. Mar. 30) Common Merganser (Depart. Apr. 6)
11   Mountain Bluebird (Depart. Oct. 16)
12   Greater White-fronted Goose (Depart. Apr. 14) Mallard (Depart. Nov. 27) Northern Pintail (Depart. Nov. 19)
13   Redhead (Depart. Nov. 9) Killdeer (Depart. Oct. 19)
16   Ring-billed Gull (Depart. May 12)
18   Canvasback (Depart. Nov. 14) Bufflehead (Depart. Apr. 21) Herring Gull (Depart. Apr. 21)
19   Lesser Scaup (Depart. May 11) Red-breasted Merganser (Depart. Apr. 20) Merlin (Depart. Mar. 19)
20   Green-winged Teal (Depart. Nov. 2) Peregrine Falcon (Depart. Sept. 22) Belted Kingfisher (Depart. Nov. 15) Fox Sparrow (Depart. Apr. 10)
21   Ring-necked Duck (Depart. Apr. 21)
22   American Wigeon (Depart. Nov. 18) Rusty Blackbird (Depart. Apr. 14)
23   Northern Shoveler (Depart. Nov. 4) Eastern Bluebird (Depart. Nov. 5)
26   Hooded Merganser (Depart. Apr. 25) Mourning Dove (Depart. Nov. 1) Common Grackle (Depart. Oct. 28)
27   Tundra Swan (Depart. *ca.* May 5)
28   Trumpeter Swan (Depart. *ca.* Nov. 15) Wood Duck (Depart. Oct. 21) Gadwall (Depart. Nov. 21)
29   Long-tailed Duck (Depart. *ca.* May 1) American Coot (Depart. Nov. 2)

April
2   Blue-winged Teal (Depart. Oct. 10) Great Blue Heron (Depart. Oct. 13)
3   Ruddy Duck (Depart. Nov. 27) McCown's Longspur (Depart. Oct. 1)
4   Loggerhead Shrike (Depart. Sept. 19)
5   Pied-billed Grebe (Depart. Nov. 4)
7   White-winged Scoter (Depart. *ca.* May 1)
8   Song Sparrow (Depart. Dec. 20) Eastern Meadowlark (Depart. Oct. 10)
10   American Woodcock (Depart. Oct. 15) Franklin's Gull (Depart. May 14) Purple Martin (Depart. Aug. 30) Brewer's Blackbird (Depart. Nov. 5)

11   Long-billed Curlew (Depart. Aug. 18)
12   Double-crested Cormorant (Depart. Oct. 23)
      Chestnut-collared Longspur (Depart. Oct. 8)
13   Greater Yellowlegs (Depart. May 5) Wilson's
      Snipe (Depart. Sept. 18) Ruby-crowned
      Kinglet (Depart. May 10)
14   Turkey Vulture (Depart. Sept. 26) Lesser Yel-
      lowlegs (Depart. May 13)
16   Horned Grebe (Depart. May 6) Eastern Phoe-
      be (Depart. Sept. 26) Say's Phoebe (Depart.
      Sept. 14)
17   Brown-headed Cowbird (Depart. Oct. 7)
18   Swainson's Hawk (Depart. Sept. 26) Vesper
      Sparrow (Depart. Oct. 9)
20   Hermit Thrush (Depart. Apr. 26) Sprague's
      Pipit (Depart. Apr. 21) Field Sparrow (De-
      part. Oct. 6)
21   Osprey (Depart. May 5) Baird's Sandpiper
      (Depart. May 13) Yellow-headed Blackbird
      (Depart. Sept. 18) Great-tailed Grackle (De-
      part. *ca.* Oct. 15)
22   Eared Grebe (Depart. Oct. 16) Eastern &
      Spotted Towhees (Depart. Oct. 15) Savannah
      Sparrow (Depart. May 10)
23   American Pipit (Depart. Apr. 28) Yellow-
      rumped Warbler (Depart. May 14) Chipping
      Sparrow (Depart. Oct. 2) Swamp Sparrow
      (Depart. Oct. 24)
24   Burrowing Owl (Depart. Sept. 16) Bewick's
      Wren (Depart. Sept. 20) House Wren (De-
      part. Sept. 26)
25   Black-crowned Night-heron (Depart. Sept. 6)
      Barn swallow (Depart. Sept. 30)
26   Cinnamon Teal (Depart. *ca.* Sept. 19) Broad-
      winged Hawk (Depart. May 15) Brown
      Thrasher (Depart. Sept. 28) Lincoln's Spar-
      row (Depart. May 13)
27   Green Heron (Depart. Sept. 18) Willet (De-
      part. Aug. 24) Chimney Swift (Depart. Oct. 7)
28   Red-throated Loon (Depart. *ca.* May 7)
      American White Pelican (Depart. Apr. 28)
      Snowy Plover (Depart. Aug. 21) American
      Avocet (Depart. Sept. 4) Semipalmated Sand-
      piper (Depart. May 15) Pectoral Sandpiper
      (Depart. May 13) Forster's Tern (Depart.
      Sept. 11) Northern Rough-winged Swallow
      (Depart. Sept. 3) Cliff Swallow (Depart. Sept.
      4)
29   Great Egret (Depart. Sept. 1) Marbled God-
      wit (Depart. May 7) White-rumped Sandpip-
      er (Depart. May 15) Tree Swallow (Depart.

Sept. 17) Baird's Sparrow (Depart. *ca.* May
5) Henslow's Sparrow (Depart. Sept. 26) Le
Conte's Sparrow (Depart. May 2)
30   Great Crested Flycatcher (Depart. Sept. 6)
      Orange-crowned Warbler (Depart. May 13)

May
1   Long-billed Dowitcher (Depart. May 11)
2   Upland Sandpiper (Depart. Aug. 20) Least
     Sandpiper (Depart. May 12) Wilson's Phala-
     rope (Depart. Sept. 8) Whip-poor-will (De-
     part. Sept. 2) Cassin's Kingbird (Depart.
     Sept. 17) Scissor-tailed Flycatcher (Depart.
     Sept. 14) Rock Wren (Depart. Oct. 27) Blue-
     gray Gnatcatcher (Depart. Sept. 1) Northern
     Mockingbird (Depart. Sept. 11)
3   American Bittern (Depart. Oct. 6) Piping
     Plover (Depart. Aug. 19) Hudsonian Godwit
     (Depart. May 15) Eastern Kingbird (Depart.
     Sept. 9) Clay-colored Sparrow (Depart. May
     16)
4   Solitary Sandpiper (Depart. May 13) Spotted
     Sandpiper (Depart. Sept. 9) Black-and-white
     Warbler (Depart. Sept. 12)
5   Least Bittern (Depart. Aug. 17) Common
     Tern (Depart. *ca.* May 11) Western Kingbird
     (Depart. Sept. 3) Marsh Wren (Depart. Oct.
     2) Northern Parula (Depart. Sept. 12) Palm
     Warbler (Depart. May 9) Brewer's Sparrow
     (Depart. Sept. 7) Lark Sparrow (Depart. Sept.
     3)
6   Western Grebe (Depart. Oct. 3) Yellow-
     crowned Night-heron (Depart. Sept. 5) King
     Rail (Depart. *ca.* Aug. 7) Sora (Depart. Sept.
     30) Sanderling (Depart. May 13) Common
     Poorwill (Depart. Sept. 4) Bank Swallow
     (Depart. Sept. 8) Swainson's Thrush (Depart.
     May 27) Grasshopper Sparrow (Depart. Sept.
     9) Baltimore & Bullock's Orioles (Depart.
     Sept. 7)
7   Common Loon (Depart. May 16) Snowy
     Egret (Depart. *ca.* Aug. 17) American Gold-
     en-plover (Depart. *ca.* May 20) Yellow-throat-
     ed Vireo (Depart. Sept. 9) Nashville Warbler
     (Depart. May 14) Yellow Warbler (Depart.
     Sept. 3) Northern Waterthrush (Depart. May
     14) Common Yellowthroat (Depart. Sep. 13)
     Rose-breasted Grosbeak (Depart. Sept. 10)
8   Little Blue Heron (Depart. *ca.* Aug. 19)
     Virginia Rail (Depart. Sept. 16) Western
     Sandpiper (Depart. May 13) Red-headed

Woodpecker (Depart. Sept. 20) Olive-sided Flycatcher (Depart. May 24) Least Flycatcher (Depart. Sept. 5) Warbling Vireo (Depart. Sept. 9) Sedge Wren (Depart. Sept. 28) Tennessee Warbler (Depart. May 28) Yellow-throated Warbler (Depart. Sept. 9) Louisiana Waterthrush (Depart. Aug. 29)

9   Cattle Egret (Depart. *ca.* Aug. 29) Blue-headed Vireo (Depart. Oct. 1) Gray-cheeked Thrush (Depart. May 27) Black-throated Green Warbler (Depart. *ca.* May 20) Orchard Oriole (Depart. Aug. 24)

10   Whimbrel (Depart. May 27) Eastern Wood-pewee (Depart. Sept. 10) White-eyed Vireo (Depart. Sept. 22) Wood Thrush (Depart. Sept. 10) Kentucky Warbler (Depart. Aug 29) Scarlet Tanager (Depart. Aug. 23) Green-tailed Towhee (Depart. Sept. 15) Lark Bunting (Depart. Aug. 30) Indigo Bunting (Depart. Aug. 28)

11   Common Moorhen (Depart. Aug. 22) Stilt Sandpiper (Depart. May 17) Gray Catbird (Depart. Sept. 24)

12   Semipalmated Plover (Depart. *ca.* June 1) Black Tern (Depart. Sept. 2) Ruby-throated Hummingbird (Depart. Sept. 13) Blue-winged Warbler (Depart. *ca.* June 1) Cape May Warbler (Depart. *ca.* May 20) Blackpoll Warbler (Depart. Sept. 9) American Redstart (Depart. Sept. 10) Prothonotary Warbler (Depart. Sept. 11) Wilson's Warbler (Depart. May 19)

13   Bell's Vireo (Depart. Sept. 8) Violet-green Swallow (Depart. Aug. 27) Ovenbird (Depart. Sept. 11)

14   Red-necked Phalarope (Depart. May 19) Red-eyed Vireo (Depart. Sept. 7) Blackburnian Warbler (Depart. May 17) Black-headed Grosbeak (Depart. Aug. 29)

15   Mississippi Kite (Depart. *ca.* Sept. 12) Acadian Flycatcher (Depart. Aug. 28) Willow Flycatcher (Depart. Sept. 2) Veery (Depart. May 18) Chestnut-sided Warbler (Depart. May 23) Magnolia Warbler (Depart. May 19) MacGillivray's Warbler (Depart. *ca.* May 25) Yellow-breasted Chat (Depart. Sept. 9) Summer Tanager (Depart. Sept. 10)

16   Black-bellied Plover (Depart. *ca.* June 1) Yellow-bellied Flycatcher (Depart. Sept. 4) Black-throated Blue Warbler (Depart. *ca.* May 20) Lazuli Bunting (Depart. Aug. 30)

Dickcissel (Depart. Aug. 22) Bobolink (Depart. Aug. 8)

17   Bay-breasted Warbler (Depart. May 19)

18   White-throated Swift (Depart. Aug. 29) Connecticut Warbler (Depart. *ca.* June 1)

19   Mourning Warbler (Depart. Sept. 8) Western Tanager (Depart. Sept. 15)

20   Canada Warbler (Depart. *ca.* June 1) Blue Grosbeak (Depart. Aug. 27)

21   Common Nighthawk (Depart. Sept. 18) Western Wood-pewee (Depart. Set. 4) Alder Flycatcher (Depart. June 7)

23   Least Tern (Depart. Aug. 14) Yellow-billed Cuckoo (Depart. Sept. 15) Philadelphia Vireo (Depart. Sept. 21)

24   Black-billed Cuckoo (Depart. Aug. 30)

June
18   Chuck-will's-widow (Depart. *ca.* Aug. 15)

Median Fall Arrival/Departure Dates

August
2   Least Sandpiper (Depart. Sept. 18)
5   Semipalmated Sandpiper (Depart. Sept. 18)
8   Long-billed Dowitcher (Depart. Oct. 14)
9   Solitary Sandpiper (Depart. Sept. 1)
11   Semipalmated Plover (Depart. Sept. 18)
12   White-rumped Sandpiper (Depart. *ca.* Oct. 4) Baird's Sandpiper (Depart. Oct. 6)
15   Lesser Yellowlegs (Depart. Oct. 5) Blue-winged Warbler (Depart. Sept. 15)
18   Greater Yellowlegs (Depart. Oct. 7)
19   Alder Flycatcher (Depart. *ca.* Sept. 10)
20   Hooded Warbler (Depart. *ca.* Sept. 12)
29   Northern Waterthrush (Depart. *ca.* Oct. 10)

September
1   Wilson's Warbler (Depart. Sept. 26) Canada Warbler (Depart. *ca.* Sept. 23)
2   Olive-sided Flycatcher (Depart. Sept. 20) Chestnut-sided Warbler (Depart. Sept. 26)
3   Blackburnian Warbler (Depart. Oct. 3)
4   Yellow-bellied Flycatcher (Depart. Sept. 4)
5   Least Flycatcher (Depart. Sept. 5)
7   Franklin's Gull (Depart. Oct. 17)
8   Swainson's Thrush (Depart. Sept. 28) Tennessee Warbler (Depart. Oct. 5) Mourning Warbler (Depart. Oct. 7) MacGillivray's Warbler (Depart. Oct. 10)
9   Magnolia Warbler (Depart. Oct. 1) Clay-

colored Sparrow (Depart. Oct. 8)
10   Nashville Warbler (Depart. Oct. 8)
11   Pectoral Sandpiper (Depart. Oct. 4)
12   Ring-billed Gull (Depart. Nov. 20)
13   Veery (Depart. *ca.* Sept. 20)
14   Bay-breasted Warbler (Depart. Sept. 22)
15   Lincoln's Sparrow (Depart. Oct. 19)
16   Sharp-shinned Hawk (Depart. Mar. 29)
18   Black-throated Green Warbler (Depart. *ca.* Oct. 2)
19   Common Tern (Depart. *ca.* Oct. 5) Orange-crowned Warbler (Depart. Oct. 15) Savannah Sparrow (Depart. Oct. 19)
23   Black-throated Blue Warbler (Depart. *ca.* Oct. 1)
24   American White Pelican (Depart. Oct. 16)
26   Townsend's Solitaire (Depart. Mar. 20) Sprague's Pipit (Depart. Oct. 23) Baird's Sparrow (Depart. Oct. 18)
28   Black-bellied Plover (Depart. Oct. 6) American Golden-Plover (Depart. Oct. 10) Yellow-rumped Warbler (Depart. Oct. 22)
30   Cape May Warbler (Depart. *ca.* Oct. 4) Connecticut Warbler (Depart. *ca.* Oct. 10) Song Sparrow (Depart. Dec. 20) Swamp Sparrow (Depart. Oct. 24)

October
1   Gray-cheeked Thrush (Depart. *ca.* Oct. 20)
2   American Pipit (Depart. Oct. 26)
3   Yellow-bellied Sapsucker (Depart. Mar. 23) White-throated Sparrow (Depart. May 12) White-crowned Sparrow (Depart. May 15)
4   Snow Goose (Depart. Dec. 2)
5   Palm Warbler (Depart. *ca.* Nov. 1)
6   Trumpeter Swan (Depart. *ca.* Nov. 15) Hermit Thrush (Depart. Oct. 16) Dark-eyed Junco (Depart. Mar. 23)
8   Horned Grebe (Depart. Nov. 11) Sandhill Crane (Depart. Nov. 5)
11   Fox Sparrow (Depart. Nov. 11)
12   Ring-necked Duck (Depart. Nov. 17)
14   Harris's Sparrow (Depart. May 12)
16   Winter Wren (Depart. Apr. 13) Pine Siskin (Depart. May 12)
18   Lesser Scaup (Depart. Nov. 17)
19   Bufflehead (Depart. Nov. 24) Golden-crowned Kinglet (Depart. Apr. 10)
21   American Tree Sparrow (Depart. Apr. 6)
23   Greater White-fronted Goose (Depart. Nov. 6)
25   Common Loon (Depart. Nov. 2) Gray-

crowned Rosy-Finch (Depart. Feb. 12)
26   Herring Gull (Depart. Nov. 28)
27   Purple Finch (Depart. Apr. 23)

November
2   Rough-legged Hawk (Depart. Mar. 26)
3   Rusty Blackbird (Depart. Dec. 26)
5   Hooded Merganser (Depart. Nov .22) Smith's Longspur (Depart. Apr. 8)
9   Northern Shrike (Depart. May 11) Evening Grosbeak (Depart. Apr. 25)
10   White-winged Scoter (Depart. *ca.* Dec. 5)
11   Snow Bunting (Depart. Feb. 10)
12   Lapland Longspur (Depart. Feb. 27) Red Crossbill (Depart. Apr. 1)
13   Common Merganser (Depart. Dec. 17)
18   Red-breasted Merganser (Depart. *ca.* Dec. 15) Red-throated Loon (Depart. *ca.* Dec. 2)
20   Bohemian Waxwing (Depart. Feb. 28)
21   Common Goldeneye (Depart. Dec. 14)
22   Tundra Swan (Depart. *ca.* Dec. 1)
24   Pine Grosbeak (Depart. Mar. 10)
26   Common Redpoll (Depart. Mar. 17)
27   Long-tailed Duck (Depart. *ca.* Dec. 15)

December
4   Snowy Owl (Depart. Feb. 5)

Regular Permanent Nebraska Residents

Canada Goose (fewer in winter)
Gray Partridge
Ring-necked Pheasant
Greater Prairie-chicken
Sharp-tailed Grouse
Wild Turkey
Northern Bobwhite
Northern Harrier (fewer in winter)
Bald Eagle (plus wintering migrants)
Cooper's Hawk
Red-shouldered Hawk
Red-tailed Hawk (plus wintering migrants)
Ferruginous Hawk (fewer in winter)
Golden Eagle (fewer in winter, in west)
American Kestrel (fewer in winter)
Prairie Falcon
Rock Pigeon
Eurasian Collared-dove
Mourning Dove (fewer in winter)
Barn Owl
Eastern Screech-owl
Great Horned Owl

Barred Owl
Long-eared Owl (fewer in winter)
Short-eared Owl (fewer in winter)
Red-bellied Woodpecker
Downy Woodpecker
Hairy Woodpecker
Northern Flicker (fewer in winter)
Pileated Woodpecker (rare, only in east)
Blue Jay
Pinyon Jay (rare, only in west)
Clark's Nutcracker (rare, only in west)
Black-billed Magpie
American Crow (plus wintering migrants)
Horned Lark (plus wintering migrants)
Carolina Wren
Black-capped Chickadee
Tufted Titmouse
White-breasted Nuthatch
Red-breasted Nuthatch
Pygmy Nuthatch
Brown Creeper
European Starling (plus wintering migrants)
House Finch
Red Crossbill
Northern Cardinal
American Goldfinch
House Sparrow

Median Spring and Fall Arrival Dates, by Taxonomic Sequence

Sample sizes for each species' median arrival date are indicated in parentheses, which also provide a rough index to the species' relative abundance during migration. Median departure dates are based on generally smaller sample sizes, which are not indicated.

Anatidae: Swans, Geese and Ducks
Greater White-fronted Goose: Mar. 12 (29) (Depart. Apr. 14); Oct. 23 (19) (Depart. Nov. 6)
Snow Goose: Mar. 9 (36) (Depart. Apr. 20); Oct. 4 (40) (Depart. Dec. 2)
Tundra Swan: Mar. 27 (20) (Depart. *ca.* May 5): Nov. 22 (11) (Depart. *ca.* Dec. 1)
Trumpeter Swan: Mar. 28 (8) (Depart. *ca.* Nov. 15); (overwinters locally, Sandhills)
Wood Duck: Mar. 28 (69) (Depart. Oct. 21)
Gadwall: Mar. 28 (48) (Depart. Nov. 21)
American Wigeon: Mar. 22 (67) (Depart. Nov. 18)
Mallard Mar. 12 (43) (Depart. Nov. 27) (often overwinters locally)

Blue-winged Teal: Apr. 2 (68) (Depart. Oct. 10)
Cinnamon Teal: Apr. 26 (62) (Depart. *ca.* Sept 19)
Northern Shoveler: Mar 23 (70) (Depart. Nov. 4)
Northern Pintail: Mar. 12 (60) (Depart. Nov. 19)
Green-winged Teal: Mar. 20 (58) (Depart. Nov. 2)
Canvasback: Mar. 18 (68) (Depart. Nov. 14)
Redhead: Mar. 13 (60) (Depart. Nov. 9)
Ring-necked Duck: Mar. 21 (42) (Depart. Apr. 21); Oct. 12 (27) (Depart. Nov. 17)
Lesser Scaup: Mar. 19 (69) (Depart. May 11); Oct. 18 (45) (Depart. Nov. 17)
White-winged Scoter: Apr. 7 (5) (Depart. *ca.* May 5); Nov. 10 (21) (Depart. *ca.* Dec. 5)
Long-tailed Duck: Mar. 29 (13) (Depart. *ca.* May 1); Nov. 27 (10) (Depart. *ca.* Dec. 15) (sometimes overwinters)
Bufflehead: Mar. 18 (53) (Depart. Apr. 21); Oct. 19 (34) (Depart. Nov. 24)
Common Goldeneye: Mar 5 (35) (Depart. Mar. 30); Nov. 21 (34) (Depart. Dec. 14) (sometimes overwinters)
Hooded Merganser: Mar. 26 (74) (Depart. Apr. 25); Nov. 5 (16) (Depart. Nov. 22)
Red -breasted Merganser: Mar. 19 (61) (Depart. Apr. 20); Nov. 18 (16) (Depart. *ca.* Dec. 15)
Common Merganser: Mar. 9 (50) (Depart. Apr. 6); Nov. 13 (38) (Depart. Dec. 17) (sometimes overwinters)
Ruddy Duck: Apr. 3 (67); (Depart. Nov. 27)

Gaviidae: Loons
Red-throated Loon: Apr. 28 (5) (Dep.*ca.* May 7); Nov. 18 (16) (Depart. *ca.* Dec. 2)
Common Loon: May 7 (55) (Depart. May 16); Oct. 24 (25) (Depart. Nov. 2)

Podicipedidae: Grebes
Pied-billed Grebe: Apr. 5 (116) (Depart. Nov. 4)
Horned Grebe: Apr. 16 (62) (Depart. May 6); Oct. 8 (17) (Depart. Nov. 11)
Eared Grebe: Apr. 22 (105) (Depart. Oct. 16)
Western Grebe: May 6 (77) (Depart. Oct. 3)

Pelecanidae: Pelicans
American White Pelican: Mar. 28 (84) (Depart. Apr. 28); Sept. 24 (28) (Depart. Oct. 16)

Phalacrocoracidae: Cormorants
Double-crested Cormorant: Apr. 12 (102) (Depart. Oct. 23)
Ardeidae: Bitterns and Herons

American Bittern: May 3 (109) (Depart. Oct. 6)
Least Bittern: May 5 (39) (Depart. Aug. 17)
Great Blue Heron: Apr. 2 (87) (Depart. Oct. 13)
Great Egret: Apr. 29 (62) (Depart. Sept. 1)
Snowy Egret: May 7 (24) (Depart. *ca.* Aug. 17)
Little Blue Heron: May 8 (35) (Depart. *ca.* Aug 19)
Cattle Egret: May 9 (21) (Depart. *ca.* Aug. 29)
Green Heron: Apr. 27 (93) (Depart. Sept. 18)
Black-crowned Night-heron: Apr. 25 (80) (Depart. Sept. 6)
Yellow-crowned Night-heron: May 6 (43) (Depart. *ca.* Sept. 5)

Cathartidae: American Vultures
Turkey Vulture: Apr. 14 (80) (Depart. Sept. 26)

Accipitridae: Kites, Hawks & Eagles
Osprey: Apr. 21 (102) (Depart. May 5); Sept. 15 (22) (Depart. Oct. 9)
Mississippi Kite: May 15 (8) (Depart. *ca.* Sept. 12)
Cooper's Hawk: Mar. 16 (164) (Depart. Oct. 30) (frequently overwinters)
Sharp-shinned Hawk: Sept. 16 (41) (Depart. Mar. 29)
Broad-winged Hawk: Apr. 26 (82) (Depart. May 15; Sept. 12 (11) (Depart. Oct. 5)
Swainson's Hawk: Apr. 18 (93) (Depart. Sept. 26)
Rough-legged Hawk: Nov. 2 (85) (Depart. Mar. 26)

Falconidae: Falcons
Merlin:. Oct. 23 (48); (Depart. Mar. 19)
Peregrine Falcon: Mar. 20 (97) (Depart. Sept. 22)

Rallidae: Rails, Gallinules, & Coots
King Rail: May 6 (9) (Depart. *ca.* Aug. 7)
Virginia Rail: May 8 (36) (Depart. Sept. 16)
Sora: May 6 (108) (Depart. Sept. 30)
Common Moorhen: May 11 (16) (Depart. Aug. 22)
American Coot: Mar. 29 (74) (Depart. Nov. 2)

Gruidae: Cranes
Sandhill Crane: Mar. 1 (57) (Depart. Apr. 7); Oct. 8 (5) (Depart. Nov. 5)

Charadriidae: Plovers
Black-bellied Plover: May 16 (66) (Depart. *ca.* June 1); Aug. 20 (13) (Depart. Oct. 6)
American Golden-Plover: May 7 (49) (Depart. *ca.* May 20); Sept. 28 (10) (Depart. Oct. 10)
Snowy Plover: Apr. 28 (8) (Depart. Aug. 21)
Semipalmated Plover: May 12 (82) (Depart. *ca.* June 1); Aug. 11 (16) (Depart. Sept. 18)
Piping Plover: May 3 (61) (Depart. Aug. 19)
Killdeer: Mar. 13 (86) (Depart. Oct. 19)

Recurvirostridae: Stilts and Avocets
American Avocet: Apr. 28 (82) (Depart. Sept. 4)

Scolopacidae: Sandpipers Phalaropes
Greater Yellowlegs: Apr. 13 (115) (Depart. May 5); Aug. 18 (32) (Depart. Oct. 7)
Lesser Yellowlegs: Apr. 14 (124) (Depart. May 13); Aug. 15 (35) (Depart. Oct. 5)
Solitary Sandpiper: May 4 (88) (Depart. May 13); Aug. 9 (36) (Depart. Sept. 1)
Willet: Apr. 27 (104) (Depart. Aug. 24)
Spotted Sandpiper: May 4 (105) (Depart. Sept. 9)
Upland Sandpiper: May 2 (102) (Depart. Aug. 20)
Whimbrel: May 10 (11) (Depart. May 27)
Long-billed Curlew: Apr. 11 (83) (Depart. Aug. 18)
Hudsonian Godwit: May 2 (69) (Depart. May 15)
Marbled Godwit: Apr. 29 (117) (Depart. May 7)
Sanderling: May 6 (56) (Depart. May 13); Aug. 20 (17) (Depart. *ca.* Oct. 4)
Semipalmated Sandpiper: Apr. 28 (89) (Depart. May 15), Aug. 5 (23) (Depart. Sept. 18)
Western Sandpiper: May 8 (41) (Depart. May 13)
Least Sandpiper: May 2 (102) (Depart. May 12). Aug. 2 (23) (Depart. Sept. 18)
White-rumped Sandpiper: Apr. 29 (100) (Depart. May 15); Aug. 12 (11) (Depart. *ca.* Oct. 4)
Pectoral Sandpiper: Apr. 28 (102) (Depart. May 13); Sept. 11 (11) (Depart. Oct. 4)
Dunlin: May 13 (48) (Depart. *ca.* June 2)
Stilt Sandpiper: May 11 (99) (Depart. May 17)
Long-billed Dowitcher: May 1 (35) (Depart. May 11); Aug. 8 (11) (Depart. Oct. 14)
Wilson's Snipe: Apr. 13 (81); (Depart. Sept. 18)
American Woodcock: Apr. 10 (13) (Depart. Oct. 15)
Wilson's Phalarope: May 2 (15) (Depart. Sept. 8)
Red-necked Phalarope: May 14 (42) (Depart. May 19); Aug. 10 (10) (Depart. Sept. 27)

Laridae: Gulls and Terns
Franklin's Gull: Apr. 10 (89) (Depart. May 14), Sept. 7 (52) (Depart. Oct. 17)
Ring-billed Gull: Mar. 16 (80) (Depart. May 12); Sept. 12 (48) (Depart. Nov. 20)
Herring Gull: Mar. 18 (47) (Depart. Apr. 21); Oct. 26 (24) (Depart. Nov. 28)
Common Tern: May 5 (65) (Depart. May 11); Sept.

19 (24) (Depart. *ca.* Oct. 5)

Forster's Tern: Apr. 28 (58); (Depart. Sept. 11)

Least Tern: May 23 (87) (Depart. Aug. 14)

Black Tern: May 12 (130) (Depart. Sept. 2)

Columbidae: Doves & Pigeons

Mourning Dove: Mar, 26 (62) (Depart. Nov. 1)

Cuculidae: Cuckoos and Anis

Black-billed Cuckoo: May 24 (163) (Depart. Aug. 30)

Yellow-billed Cuckoo: May 23 (170) (Depart. Sept. 15)

Tytonidae: Barn Owls

Snowy Owl: Dec. 4 (18) (Depart. Feb. 5)

Burrowing Owl: Apr. 24 (119) (Depart. Sept. 16)

Caprimulgidae: Goatsuckers

Common Nighthawk: May 21 (170) (Depart. Sept. 18)

Common Poorwill: May 6 (38) (Depart. Sept. 4)

Chuck-will's-widow: June 3 (14) (Depart. *ca.* Aug. 15)

Whip-poor-will: May 2 (34) (Depart. Sept. 2)

Apodidae: Swifts

Chimney Swift: Apr. 27 (129) (Depart. Oct. 7)

White-throated Swift May 18 (26) (Depart. Aug. 29)

Trochilidae: Hummingbirds

Ruby-throated Hummingbird: May 12 (160) (Depart. Sept. 13)

Alcedinidae: Kingfishers

Belted Kingfisher: Mar. 20 (43) (Depart. Nov. 15)

Picidae: Woodpeckers

Red-headed Woodpecker: May 7 (98) (Depart. Sept. 20)

Yellow-bellied Sapsucker: Oct. 3 (34); (Depart. Mar. 23)

Tyrannidae: Tyrant Flycatchers

Olive-sided Flycatcher: May 8 (68) (Depart. May 24); Sept. 2 (18) (Depart. Sept. 20)

Western Wood-pewee: May 21 (64) (Depart. Sept. 4)

Eastern Wood-pewee: May 10 (77) (Depart. Sept. 10)

Yellow-bellied Flycatcher: May 16 (26) (Depart. Sept. 4)

Acadian Flycatcher: May 15 (55) (Depart. Aug. 28)

Alder Flycatcher: May 21 (8) (Depart. June 7); Aug 19 (5) (Depart. *ca.* Sept. 10)

Willow Flycatcher: May 15 (78) (Depart. Sept. 2)

Least Flycatcher: May 8 (100) (Depart. Sept. 5)

Eastern Phoebe: Apr. 16 (169) (Depart. Sept. 26)

Say's Phoebe: Apr. 16 (129) (Depart. Sept. 14)

Great Crested Flycatcher: Apr. 30 (130) (Depart. Sept. 6)

Cassin's Kingbird: May 2 (18) (Depart. Sept. 17)

Western Kingbird: May 5 (117) (Depart. Sept. 3)

Eastern Kingbird: May 3 (73) (Depart. Sept. 9)

Scissor-tailed Flycatcher: May 2 (17) (Depart. Sept. 14)

Laniidae: Shrikes

Northern Shrike: Nov. 9 (44) (Depart. May 11)

Loggerhead Shrike: Apr. 4 (95) (Depart. Sept. 19)

Vireonidae: Vireos

White-eyed Vireo: May 10 (44) (Depart. Sept. 22)

Bell's Vireo: May 13 (114) (Depart. Sept. 8)

Blue-headed Vireo: May 9 (77) (Depart. Oct. 1)

Yellow-throated Vireo: May 7 (80) (Depart. Sept. 9)

Warbling Vireo: May 8 (112) (Depart. Sept. 9)

Philadelphia Vireo: May 23 (52) (Depart. Sept. 21)

Red-eyed Vireo: May 14 (129) (Depart. Sept. 7)

Hirundinidae: Swallows

Purple Martin: Apr. 10 (143) (Depart. Aug. 30)

Tree Swallow: Apr. 29 (86) (Depart. Sept. 17)

Violet-green Swallow: May 13 (38) (Depart. Aug. 27)

Northern Rough-winged Swallow: Apr. 28 (136) (Depart. Sept. 3)

Bank Swallow: May 6 (104) (Depart. Sept. 8)

Cliff Swallow: Apr. 28 (15) (Depart. Sept. 4)

Barn Swallow: Apr. 25 (155) (Depart. Sept. 30)

Troglodytidae: Wrens

Rock Wren: May 2 (83) (Depart. Oct. 27)

Bewick's Wren: Apr. 24 (44) (Depart. Sept. 20)

House Wren: Apr. 24 (136) (Depart. Sept 26)

Winter Wren: Oct. 16 (38) (Depart. Apr. 13)

Sedge Wren: May 8 (25) (Depart. Sept. 28)

Marsh Wren: May 5 (78) (Depart. Oct. 2)

Regulidae: Kinglets

Golden-crowned Kinglet: Oct. 19 (75) (Depart. Apr. 10)

Ruby-crowned Kinglet: Apr. 13 (74) (Depart. May 10); Sept. 23 (75) (Depart. Oct. 28)

Sylviidae: Gnatcatchers
Blue-gray Gnatcatcher: May 2 (85) (Depart. Sept. 1)

Turdidae: Thrushes and Allies
Eastern Bluebird: Mar. 23 (123) (Depart. Nov. 5)

Mountain Bluebird: Mar. 11 (84) (Depart. Oct. 16)

Townsend's Solitaire: Sept. 26 (50) (Depart. Mar. 20)

Veery: May 15 (108) (Depart. May 18); Sept. 13 (7) (Depart. *ca.* Sept. 20)

Gray-cheeked Thrush: May 9 (100) (Depart. May 27); Oct. 1 (5) (Depart. *ca.* Oct. 20)

Swainson's Thrush: May 6 (141) (Depart. May 27); Sept. 8 (51) (Depart. Sept. 28)

Hermit Thrush: Apr. 20 (94) (Depart. Apr. 26) Oct. 6 (14) (Depart. Oct. 16)

Wood Thrush: May 10 (120) (Depart. Sept. 10)

American Robin: Feb. 20 (45) (Depart. Nov. 19)

Mimidae: Mockingbirds, Thrashers, etc.
Gray Catbird: May 11 (134) (Depart. Sept. 24)

Northern Mockingbird; May 2 (132) (Depart. Sept. 11)

Brown Thrasher: Apr. 26 (134) (Depart. Sept. 28)

Motacillidae: Pipits
American Pipit: Apr. 23 (125) (Depart. Apr. 28); Oct. 2 (18) (Depart. Oct. 26)

Sprague's Pipit: Apr. 20 (41) (Depart. Apr. 21); Sept. 26 (17) (Depart. Oct. 23)

Bombycillidae: Waxwings
Bohemian Waxwing: Nov. 20 (11) (Depart. Feb. 28)

Cedar Waxwing: Feb. 24 (54) (Depart. Oct. 4)

Parulidae: Wood Warblers
Blue-winged Warbler: May 12 (22) (Dept. *ca.* June 1); Aug. 15 (4) (Depart. *ca.* Sept. 15)

Tennessee Warbler: May 8 (95) (Depart. May 28); Sept. 8 (31) (Depart. Oct. 5)

Orange-crowned Warbler: Apr. 30 (112) (Depart. May 13); Sept. 19 (61) (Depart. Oct. 15)

Nashville Warbler: May 7 (81) (Depart. May 14); Sept. 10 (41) (Depart. Oct. 8)

Northern Parula: May 5 (34) (Depart. Sept. 12)

Yellow Warbler: May 7 (126) (Depart. Sept. 3)

Chestnut-sided Warbler: May 15 (61) (Depart. May 23); Sept. 2 (6) (Depart. Sept. 26)

Magnolia Warbler: May 15 (121) (Depart. May 19); Sept. 9 (13) (Depart. Oct. 1)

Cape May Warbler: May 12 (14) (Depart. *ca.* May 20); Sept. 30 (5) (Depart. *ca.* Oct. 4)

Black-throated Blue Warbler: May 16 (9) (Depart. *ca.* May 20); Sept. 23 (27) (Depart. *ca.* Oct. 1)

Yellow-rumped Warbler: Apr. 23 (75) (Depart. May 14); Sept. 28 (80) (Depart. Oct. 22)

Black-throated Green Warbler: May 9 (44) (Depart. *ca.* May 20); Sept. 18 (16) (Depart. *ca.* Oct. 2)

Blackburnian Warbler: May 14 (76) (Depart. May 17); Sept. 3 (10) (Depart. Oct. 3)

Yellow-throated Warbler: May 8 (21) (Depart. Sept. 9)

Palm Warbler: May 5 (63) (Depart. May 9); Oct. 5 (10) (Depart. *ca.* Nov. 1)

Bay-breasted Warbler: May 17 (41) (Depart. May 19); Sept. 14: (7) (Depart. Sept. 22)

Blackpoll Warbler: May 12 (120) (Depart. Sept. 9)

Cerulean Warbler: May 14 (38) (Depart. Aug. 15)

Black-and-white Warbler: May 4 (92) (Depart. Sept. 12)

American Redstart: May 12 (131) (Depart. Sept. 10)

Prothonotary Warbler: May 12 (32) (Depart. Sept. 11)

Ovenbird: May 13 (130) (Depart. Sept. 11)

Northern Waterthrush: May 7 (135 (Depart. May 14); Aug. 29 (8) (Depart. *ca.* Oct. 10)

Louisiana Waterthrush: May 8 (76) (Depart. Aug. 29)

Kentucky Warbler: May 10 (40) (Depart. Aug. 29)

Connecticut Warbler: May 18 (20) (Depart. *ca.* June 1); Sept. 30 (10) (Depart. *ca.* Oct. 10)

Mourning Warbler: May 19 (87) (Dept. May 28); Sept. 8 (18) (Depart. Oct. 7)

MacGillivray's Warbler: May 15 (98) (Dep. *ca.* May 25); Sept. 8 (18) (Depart. *ca.* Oct. 10)

Common Yellowthroat: May 7 (107) (Depart. Sept. 13)

Hooded Warbler May 11 (27) (Depart. *ca.* May 25); Aug. 20 (5) (Depart. *ca.* Sept. 12)

Wilson's Warbler: May 12 (101) (Depart. May 19); Sept. 1 (69) (Depart. Sept. 26)

Canada Warbler: May 20 (28) (Depart. *ca.* June 1); Sept. 1 (14) (Depart. *ca.* Sept 23)

Yellow-breasted Chat: May 15 (120) (Depart. Sept. 9)

Thraupidae: Tanagers
Summer Tanager: May 15 (29) (Depart. Sept. 10)
Scarlet Tanager: May 10 (132) (Depart. Aug. 23)
Western Tanager: May 19 (63) (Depart. Sept. 15)

Emberizidae: Towhees & Sparrows
Eastern & Spotted Towhee; Apr. 22 (69) (Depart. Oct. 15) (some may overwinter)
Green-tailed Towhee: May 10 (19) (Depart. Sept. 15)
American Tree Sparrow: Oct. 21 (127) (Depart. Apr. 6)
Chipping Sparrow: Apr. 23 (100) (Depart. Oct 2)
Clay-colored Sparrow: May 3 (129) (Depart. May 16); Sept. 9 (41) (Depart. Oct. 8)
Brewer's Sparrow: May 5 (27) (Depart. Sept. 7)
Field Sparrow: Apr. 20 (81) (Depart. Oct. 6)
Vesper Sparrow: Apr. 18 (30) (Depart. Oct. 9)
Lark Sparrow: May 5 (125) (Depart. Sept. 3)
Lark Bunting: May 10 (104) (Depart. Aug. 30)
Savannah Sparrow: Apr. 22 (69) (Depart. May 10); Sept. 19 (39) (Depart. Oct. 19)
Baird's Sparrow: Apr. 29 (44) (Depart. ca. May 5); Sept. 26 (15) (Depart. Oct. 18)
Grasshopper Sparrow: May 6 (85) (Depart. Sept. 9)
Henslow's Sparrow: Apr. 29 (21) (Depart. Sept. 26)
Le Conte's Sparrow: Apr. 29 (54) (Depart. May 2); Sept. 22 (21) (Depart. Oct. 20)
Fox Sparrow: Mar. 20 (53) (Depart. Apr. 10); Oct. 11 (31) (Depart. Nov. 11)
Song Sparrow: Apr. 8 (45) (Depart. Dec. 20)
Lincoln's Sparrow: Apr. 26 (94) (Depart. May 13); Sept. 15 (48) (Depart. Oct. 19)
Swamp Sparrow: Apr. 23 (33) (Depart. Oct. 24)
White-throated Sparrow: Oct. 3 (65) (Depart. May 12)
Harris's Sparrow: Oct. 14 (115) (Depart. May 12)
White-crowned Sparrow: Oct. 3 (98) (Depart. May 15)
Dark-eyed Junco: Oct. 6 (105) (Depart. Mar. 23)
McCown's Longspur: Apr. 3 (26) (Depart. Oct. 1)
Lapland Longspur: Nov. 12 (56) (Depart. Feb. 27)
Smith's Longspur: Nov. 5 (10) (Depart. Apr. 8)
Chestnut-collared Longspur: Apr. 12 (30) (Depart. Oct. 8)
Snow Bunting: Nov. 11 (11) (Depart. Feb. 10)

Cardinalidae: Cardinals & Grosbeaks
Rose-breasted Grosbeak: May 7 (134) (Depart. Sept. 10)
Black-headed Grosbeak: May 14 (114) (Depart. Aug. 29)
Blue Grosbeak: May 20 (129) (Depart. Aug. 27)
Lazuli Bunting: May 16 (113) (Depart. Aug. 25)
Indigo Bunting: May 10 (99) (Depart. Aug. 28)
Dickcissel: May 16 (199) (Depart. Aug. 22)

Icteridae: Blackbirds, Orioles, etc
Bobolink: May 16 (116) (Depart. Aug. 8)
Red-winged Blackbird: Mar. 3 (90) (Depart. Nov. 21)
Eastern Meadowlark: Apr. 8 (59) (Depart. Oct. 10 (sometimes overwinters)
Western Meadowlark: Mar. 4 (61) (Depart. Oct. 28 (sometimes overwinters)
Yellow-headed Blackbird: Apr. 21 (103) (Depart. Sept. 18)
Rusty Blackbird: Mar. 22 (45) (Depart. Apr. 14); Nov. 3 (25) (Depart. Dec. 26)
Brewer's Blackbird: Apr. 10 (63) (Depart. Nov. 5)
Common Grackle: Mar. 26 (82) (Depart. Oct. 28)
Great-tailed Grackle: Apr. 21 (5) (Depart. *ca.* Oct. 15)
Brown-headed Cowbird: Apr. 17 (83) (Depart. Oct. 7)
Orchard Oriole: May 9 (188) (Depart. Aug. 24)
Baltimore & Bullock's Orioles: May 6 (192) (Depart. Sept. 7)

Fringillidae: Finches
Gray-crowned Rosy-Finch: Oct. 25 (6) (Depart. Feb. 12)
Pine Grosbeak: Nov. 24 (14) (Depart. Mar. 10)
Purple Finch: Oct. 27 (37) (Depart. Apr. 23)
Red Crossbill: Nov. 12 (31) (Depart. Apr. 1)
Common Redpoll: Nov. 26 (20) (Depart. Mar. 17)
Pine Siskin: Oct. 16 (60) (Depart. May 12)
Evening Grosbeak: Nov. 9 (34) (Depart. Apr. 25)

Some Nebraska Bird Specialties:
Where and When to See Them

Snow Goose: Missouri Valley and central Platte Valley and Rainwater Basin wetlands (March); DeSoto Bend Natl. Wildlife Refuge (late October and November)

Greater White-fronted Goose: Central Platte Valley and Rainwater Basin wetlands; March

Trumpeter Swan: Sandhills lakes (Cherry and Sheridan counties), spring to fall; Snake Creek & Blue Creek (winter)

Cackling Goose: Central Platte Valley, March, October-November; borrow pits lakes along I-80.

Cinnamon Teal: Alkaline marshes, Morrill and Scotts Bluff counties; spring to fall

Ruddy Duck: Deeper Sandhills marshes (Crescent Lake and Valentine Natl. Wildlife Refuges); spring and summer

Sharp-tailed Grouse: Sandhills and western Nebraska's mixed grasslands; especially March-April (courtship period), including Halsey National Forest and Ft. Niobrara NWR

Greater Prairie-chicken: East-central and southeast Nebraska; taller grasslands, especially March-April (courtship period), including Valentine NWR

Eared Grebe: Western Sandhills marshes; spring and summer, especially around Lakeside.

Western and Clark's Grebes: Sandhills marshes (spring, summer), Lake McConaughy (fall)

American Bittern: Sandhills marshes; spring and summer

Least Bittern: Southeastern Nebraska; marshes and sod farms

Black-crowned Night-heron: Sandhills marshes; spring and summer

White-faced Ibis: Marshes and wet meadows; Crescent Lake Natl. Wildlife Refuge, Lakeside; spring and summer

Bald Eagle: Platte and Missouri rivers, many larger reservoirs; fall to spring, plus local breeding along larger rivers and reservoirs

Swainson's Hawk: Widespread (especially Panhandle and Sandhills); spring, summer and fall

Ferruginous Hawk: Shortgrass plains, Oglala National Grassland; spring through fall

Golden Eagle: Western Nebraska, especially Panhandle; year round

Prairie Falcon: Shortgrass plains; Oglala National Grassland (most of year); Niobrara Valley (winter)

Sandhill Crane: Platte Valley cornfields, meadows and wetlands; mid-February to mid-April

Whooping Crane: Platte Valley wetlands; late March and April

American Avocet: Alkaline marshes, western Sandhills; Rainwater Basin, spring and summer

Black-necked Stilt: Alkaline marshes, Sheridan Co. and Crescent Lake Natl. Wildlife Refuge; spring and summer, also central Sandhills marshes around Lakeside.

Piping Plover: Platte, Missouri and lower Niobrara rivers; sandy islands and bars; spring and summer

Mountain Plover: Shortgrass prairie, Kimball Co.; late spring and summer

Upland Sandpiper: Sandhills meadows; spring and summer

Buff-breasted Sandpiper: Meadows, sod farms and fields; Rainwater Basin, York County; early May

Long-billed Curlew: Sandhills meadows; spring and summer

Wilson's Phalarope: Alkaline Sandhills marshes

(Sheridan County); spring and summer

Least Tern: Platte, Missouri, lower Niobrara rivers, bare sandbars and islands; spring and summer

Burrowing Owl: Panhandle (especially Morrill County), Kiowa W.M.A., Fort Niobrara Natl. Wildlife Refuge; prairie dog colonies, spring and summer

Common Poorwill: Panhandle (Pine Ridge and Wildcat Hills); rocky scrub; summer

White-throated Swift: Pine Ridge, Scotts Bluff, Wildcat Hills, other escarpments; summer

Pileated Woodpecker: Fontenelle Forest, Indian Cave State Park; mature hardwoods; year round

Cassin's Kingbird: Panhandle; open pine forests, summer and fall

Pinyon Jay: Scrubby conifers, Pine Ridge; year round

Violet-green Swallow: Bluffs, escarpments, Pine Ridge, Scotts Bluff, Wildcat Hills; summer

Rock Wren: Panhandle, rocky and badlands areas; spring and summer

Pygmy Nuthatch: Pine Ridge and Wildcat Hills ponderosa pine forests, year round

Townsend's Solitaire: Pine Ridge conifers (spring, summer); western Platte Valley juniper woods (fall, winter)

Sage Thrasher: Northwestern Panhandle; sagebrush flats, spring and fall (rare)

Black-and-white Warbler: Mature hardwood forests Niobrara Valley; spring and summer

Ovenbird: Mature hardwood forests, Niobrara & Missouri Valley; spring and summer

American Redstart: Mature hardwood forests, Niobrara & Missouri valleys; spring and summer,

Kentucky Warbler: Southeast Nebraska upland forests, Platte River State Park; spring, summer

Summer Tanager: Southeast Nebraska, mature forests Platte River State Park, Indian Cave, Schramm; spring, summer

Scarlet Tanager: Mature Missouri Valley riparian forests; spring and summer

Western Tanager: Panhandle coniferous forests; spring and summer

Brewer's Sparrow: Sandsage scrub, Panhandle, around Toadstool Park; spring and summer

Cassin's Sparrow: sandsage, Dundy, Hayes, Chase and Hitchcock counties; spring and summer

Lark Bunting: Western mixed and shortgrass prairies; spring and summer

Harris' Sparrow: Brushy edges, plum thickets, woodpiles, central and eastern Nebraska; fall to spring

McCown's Longspur: Shortgrass prairies; Panhandle, spring and summer

Chestnut-collared Longspur: Northwest and north-central Nebraska; mixed-grass prairies, spring and summer

Dickcissel: Eastern tallgrass prairies, irrigated alfalfa fields farther west; spring and summer

Bobolink: Wet meadows; especially Sandhills lake areas; spring and summer

Red Crossbill: Pine Ridge (Monroe and Sowbelly canyons), Wildcat Hills; year round

Greater Prairie-chickens

Nebraska Habitat Types of Importance to Birds*

Wooded Habitats

Forest (canopies higher than 5 m. and over 60% canopy cover)
  Lowland (Floodplain) Deciduous Forest
  Upland Coniferous Forest
  Upland Deciduous Forest

Woodland (canopies higher than 5 m. and 25-60% canopy cover)
  Lowland (Floodplain) Deciduous Woodland
  Upland Coniferous Woodland
  Upland Deciduous Woodland

Shrubland (shrubs under 5 m.; more than 25% canopy cover)
  Lowland (Floodplain) Shrubland
  Upland Shrubland
  Sandsage Shrubland/Grassland

Grasslands (Habitats dominated by non-woody plants; canopy cover over 25 %)

  Lowland Tallgrass Prairie
  Upland Tallgrass Prairie
  Sand Hills Prairie
  Mixed-grass Prairie
  Shortgrass Prairie

Sparse Vegetation (Plant cover <25%, highly variable topography)

  Badlands (Steep, eroded slopes)
  Dry Cliffs/Rock Cavities (Very steep rock escarpments)
  Rock Outcrop (Moderate to fairly steep rock escarpments)
  Sand or Gravel Flats (Barren riverine/lacustrine edges & bars)

Wetlands (Aquatic and semi-aquatic habitats)

  Alkaline (Saline) Wetlands
  Playa (Seasonal) Wetlands
  Wet Meadow/Marsh
  Open Water, Lakes & Reservoirs
  Open Water, Streams & Rivers
  Swamps/Wooded Backwaters (Woody oxbows, flooded trees)
  Open Shorelines (Lightly vegetated shores)

Other Habitats (Variable plant life-form and cover extent)

  Cropland
  Prairie Dog Town
  Urban/Parks/Bridges/Other human constructions
  Woodland Edge (Woodlands edged by shrub & herbaceous communities)

* Adapted from Steinauer, G., and S. Rolfsmeier, 2003. *Terrestrial Natural Communities of Nebraska* (Version III). Nebraska Game & Parks Comm., Lincoln, NE. 162 pp.

See Johnsgard and Dinan (2005) for habitat assignments of all regularly occurring Nebraska birds (www. nebraskabirds.org).

Latitude / Longitude Data for State-owned Birding Sites

A

Alexandria SRA   40.23253/-97.33354

Alexandria WMA 40.23614/ -97.35016

American Game Marsh WMA 42.3078/-100.06539

Anderson Bridge WMA 42.78656/ -100.93496

Arbor Lake WMA 40.90377/-96.67886

Arbor Lodge SHP 40.68059/-95.87837

Arcadia Diversion WMA 41.49138/-99.23107

Arnold Lake SRA 41.41351/-100.19721

Arrowhead WMA 40.10126/ -96.8716

Ash Grove WMA 40.03491/-98.96852

Ash Hollow SHP 41.29658/-102.11787

Ashfall SHP 42.42188/-98.15765

Aspinwall Bend WMA 40.31864/ -95.66079

Atkinson SRA 42.5386/ -99.00032

B

Ballards Marsh WMA 42.59624/-100.54873

Bartley Diversion Dam WMA 40.2263/-100.38271

Bassway Strip WMA 40.68616/-98.94346

Basswood Ridge WMA 42.35145/-96.50913

Bazile Creek WMA 42.80666/-97.8964

Beaver Bend WMA 41.58544/ -97.88427

Big Alkali Lake WMA 42.63914/ -100.60673

Big Springs WMA 41.05397/-102.0535

Birdwood Lake WMA 41.11637/-100.83425

Bittern's Call WMA 40.84145/ -99.84629

Bittersweet WMA 41.04368/-102.13962

Black Island WMA 42.00286/-96.9979

Blue Bluffs WMA 40.7401/ -97.02459

Blue Heron WMA 40.91495/-100.17323

Blue Hole East WMA 40.68291/-99.32657

Blue Hole WMA 40.6856/-99.37854

Blue River SRA 40.70627/-97.11971

Bluebill WMA 40.63667/-97.70302

Bluestem 40.63915/-96.80171

Bluewater Battlefield SHP 41.39351/-102.18555

Bluewing WMA 40.36524/-98.04227

Bobcat WMA 42.72726/-99.87732

Bohemia Prairie WMA 42.6761/-98.12892

Bordeaux WMA 42.80069/-102.90105

Borman Bridge WMA 42.84717/-100.51532

Bowman Lake SRA 41.27632/-98.99176

Bowring Ranch SHP 42.95805/-101.67211

Bowwood WMA 40.16763/ -96.24152

Box Butte Reservoir SRA 42.45567/-103.11284

Box Elder Canyon WMA 41.02524/-100.57254

Brady WMA 40.9989/-100.37144

Bramble WMA 41.42486/-96.689

Branched Oak SRA 40.97945/-96.87038

Branched Oak WMA 40.97638/-96.84644

Bridgeport SRA 41.68015/-103.11064

Brownville SRA 40.39454/-95.65057

Buckskin Hills WMA 42.626/-96.92536

Buffalo Bill SRA 41.16884/-100.78402

Buffalo Bill's Ranch SHP 41.16219/-100.79494

Buffalo Creek WMA 41.69498/-103.59

Bufflehead WMA 40.67194/-99.01571

Bulrush WMA 40.39223/-98.07566

Bur Oak WMA 40.89722/-96.99995

Burchard Lake WMA 40.16796/-96.30

C

Calamus Reservoir SRA and WMA 41.87433/ -99.28

Catfish Run WMA 41.05477/-96.33596

Cattail WMA 40.775/-98.54838

Cedar Canyon WMA 41.75231/-103.77989

Cedar Creek Island WMA 41.05977/-96.06187

Cedar Valley WMA 40.72875/ -100.6971

Chadron SP 42.70914/-103.01974

Chalkrock WMA 42.79762/-97.37711

Champion Lake SRA 40.47155/-101.75244

Champion Mill SHP 40.47112/ -101.75068

Chester Island WMA 40.99075/-100.391

Chet and Jane Fleisbach WMA 41.6922/ -103.2633

Cheyenne SRA 40.76237/-98.59287

Clear Creek WMA 41.29125/-102.02603

Conestoga Lake SRA 40.76431/-96.86019

Coot Shallows WMA 40.68316/ -99.29098

Cornhusker WMA 40.90231/-98.45796

Cottonwood Canyon WMA 41.00219/-100.5

Cottonwood Lake SRA 42.91593/-101.675

Cottonwood/Steverson WMA 42.41736/-101.69

Council Creek WMA 41.43427/-97.86414

Cozad WMA 40.83674/-99.98128

Crystal Lake SRA 40.45544/-98.43981

D

Darr Strip WMA 40.80952/-99.89854

Darr WMA 40.78458/-99.85264

Davis Creek Reservoir WMA 41.42752/-98.76502

De Fair Lake WMA 41.96298/-101.71131

Dead Timber SRA 41.71698/-96.69138

Deep Well WMA 40.845/-98.22075

Denman Island WMA 40.71504/-98.72587

Diamond Lake WMA 40.04008/-96.86897

Divoky Acres WMA 40.5205/-97.1459

DLD SRA 40.58189/-98.28979

Dogwood WMA 40.69897/-99.62684

Donald Whitney Memorial WMA 40.04698/-96.86

Dry Creek WMA 42.43339/-98.60051

Dry Sandy WMA 40.34729/-97.51508

E

East Cozad WMA 40.83137/-99.95514

East Darr WMA 40.77305/-99.83367

East Gothenburg WMA 40.88945/ -100.1049

East Hershey WMA 41.12507/-100.8999

East Odessa WMA 40.66866/-99.16611

East Sutherland WMA 41.14051/-101.05509

East Willow Island WMA 40.8627/-100.03634

Elk Point Bent WMA 42.64633/-96.71479

Elkhorn WMA 42.0589/-97.42889

Elm Creek WMA 40.12343/ -98.44595

Elwood Reservoir WMA 40.62673/-99.85837

Enders Reservoir SRA 40.4334/-101.51936

Enders Reservoir WMA 40.42899/-101.56682

Eugene T. Mahoney SP 41.02601/-96.31223

F

Father Hupp WMA 40.34011/-97.61765

Ferry Landing WMA 42.76252/-97.9908

Flat Water Landing WMA 41.4008/-97.36802

Flathead WMA 40.12448/-97.18211

Flatsedge WMA 41.22293/-97.47859

Fort Atkinson SHP 41.45498/-96.01189

Fort Hartsuff SHP 41.7235/-99.02355

Fort Kearny SHP 40.6429/-99.00591

Fort Kearny SRA 40.65493/-98.99636

Fort Robinson SP 42.68664/ -103.49412

Four Mile Creek WMA 40.04057/-95.9353

Fred Thomas WMA 42.71845/-99.58421

Fremont Lakes SRA 41.44169/-96.55833

Fremont Slough WMA 41.09704/-100.66669

Frenchman WMA 40.35542/-101.09561

Frye Lake WMA 42.02011/-101.74465

G

Gadwall WMA 40.94006/ -98.03742

Gallagher Canyon SRA 40.73485/-99.97898

George D. Syas WMA 41.43304/-97.6818

Gilbert-Baker WMA 42.75739/-103.93339

Goldeneye WMA 41.03693/-102.12459

Golderod WMA 41.05883/-102.4238

Goose Lake WMA 42.11046/-98.5641

Green Wing WMA 40.44298/-97.82895

Greenhead WMA 40.44405/-97.94042

Greenvale WMA 42.54032/-98.2204

Grove Lake WMA 42.35137/-98.10155

H

Hackberry Creek WMA 42.17629/-98.15995
Hamburg Bend WMA 40.59315/ -95.76712
Hansen Memorial Reserve WMA 40.72364/-100.5314
Harold W. Andersen WMA 41.14279/-98.50286
Hayes Center WMA 40.59241/-100.93071
Hedgefield WMA 40.602/-96.56767
Hershey WMA 41.13547/-100.98802
Hickory Ridge WMA 40.3133/-96.35556
Hidden Marsh WMA 40.70951/-97.48761
High Basin WMA 40.56492/-99.63964
Holt Creek WMA 42.95987/-99.70706
Hord Lake SRA 41.10563/-97.95807
Hull Lake WMA 42.86698/-98.88138

I

Indian Cave SP 40.264/ -95.56964
Indian Creek WMA 40.06326/-98.52348

J

Jack Sinn Memorial WMA 41.04683/-96.57116
Jeffrey Canyon WMA 40.95169-100.40195
Johnson Lake SRA 40.68463/-99.82957

K

Kea Lake WMA 40.66854/-99.09076
Kea West WMA 40.6716/-99.10677
Keller Park SRA 42.66909/-99.77515
Keller Park WMA 42.65846/-99.78476
Kent Diversion Dam WMA 41.75981/ -99.26788
Killdeer WMA 40.67681/-96.76599
Kinter's Ford WMA 40.05681/-95.99797
Kiowa WMA 41.92198/-103.93851
Kirkpatrick Basin North WMA 40.82708/-97.66
Kirkpatrick Basin South WMA 40.80464/-97.72488
Kissinger Basin WMA 40.44492/-98.09886

L

Lake McConaughy SRA and WMA 41.24813/-101
Lake Minatare SRA 41.9332/-103.49549
Lake Ogallala SRA 41.22366/-101.66264

Larkspur WMA 41.08614/ -96.90367
Lewis and Clark Lake SRA 42.84337/-97.52087
Limestone Bluffs WMA 40.0096/ -98.8863
Little Blue WMA 40.15799/-97.5198
Loch Linda WMA 40.81501/-98.42541
Long Lake SRA 42.29292/-100.10351
Long Pine SRA 42.54651/-99.7109
Long Pine WMA 42.55361/-99.70319
Lookingglass Creek WMA 41.46418/-97.60173
Lores Branch WMA 40.0405/-96.08758
Louisville Boat Access WMA 41.01539/-96.1576
Louisville SRA 41.00583/-96.17064
Loup Bottoms WMA 41.36365 -98.61847
Loup Junction WMA 41.2694/ -98.4093
Loup Public Power District WMA 41.40776/ -97.78

M

Maloney Reservoir SRA 41.04836/-100.79953
Margrave WMA 40.01459/-95.46832
Marsh Hawk WMA 40.63677/-97.72174
Marsh Wren WMA 41.2685/-98.52254
Martin's Reach WMA 40.7358/ -98.6405
Mayberry WMA 40.22145/-96.33046
Medicine Creek SRA 40.37919/-100.21197
Medicine Creek WMA 40.43735/-100.26531
Memphis Lake SRA 41.10326/ -96.4456
Meridian WMA 40.20321/-97.40511
Merritt Reservoir SRA 42.60175/-100.88588
Merritt Reservoir WMA 42.60838/-100.89559
Metcalf WMA 42.83508/-102.68204
Middle Decatur Bend WMA 42.0032/-96.20486
Milburn Dam WMA 41.75779/-99.77695
Mormon Island SRA 40.82461/-98.37077
Mulberry Bend WMA 42.71541/-96.94553
Muskrat Run WMA 41.19618/-100.89433
Myrtle E. Hall WMA 41.7431/ -99.53615

N

Narrows WMA 40.08225/-98.60281
Nine Mile Creek WMA 41.90042/ -103.43712
Niobrara SP 42.74998/ -98.06724

North Lake Basin WMA 40.91502/ -97.34216

North Loup SRA 41.26428/-98.45198

North Platte FH 41.09141/-100.75723

North River WMA 41.20194/ -100.98341

Northeast Sacramento WMA 40.42482/-98.99193

O

Oak Glen WMA 40.97723/ -96.98717

Oak Valley WMA 41.95285/-97.62501

Olive Creek SRA 40.57679/-96.84611

Oliver Reservoir SRA 41.22453/ -103.82729

Omadi Bend WMA 42.35153/ -96.43272

Osage WMA 40.43056/ -96.22491

Overton WMA 40.69165/-99.54453

Oxford WMA 40.25585/-99.72345

P

Parshall Bridge WMA 42.84811/-98.84746

Pawnee Lake SRA 40.84961/ -96.88159

Pawnee Prairie WMA 40.03365/ -96.32675

Pawnee Slough WMA 41.08124/-100.53897

Pelican Point SRA 41.83331/-96.11269

Peru Boat Ramp WMA 40.48416/ -95.69842

Peru Bottoms WMA 40.49915/ -95.71448

Petersen WMA 42.6437/-103.55867

Pibel Lake SRA 41.75817/-98.53155

Pine Glen WMA 42.67304/ -99.69679

Pintail WMA 40.78544/-97.95429

Pioneer SRA 41.11189/-96.62492

Platte River SP 40.9926/ -96.21073

Platte WMA 41.09862/ -100.65424

Plum Creek Valley WMA 42.54543/-100.10433

Plum Creek WMA 40.69658/ -99.91313

Ponca SP 42.60962/ -96.71727

Ponderosa WMA 42.64095/-103.30158

Powder Horn WMA 41.71208/-96.69801

Prairie Knoll WMA 40.06227/-96.07215

Prairie Marsh West WMA 40.34539/-97.64625

Prairie Wolf WMA 41.41465/ -97.73921

Pressey WMA 41.18373/-99.70866

R

Rakes Creek WMA 40.86952/ -95.85883

Randall W. Schilling WMA 41.03363/ -95.87493

Rat and Beaver Lake WMA 42.45961/ -100.65836

Red Fox WMA 41.98952/ -97.05207

Red Willow Diversion Dam WMA 40.28104/ -100.542

Red Willow Reservoir SRA 40.35979/-100.66351

Red Willow Reservoir WMA 40.37234/ -100.71114

Red Wing WMA 42.15268/-98.10664

Redbird 42.7559/-98.43152

Redhead WMA 40.43393/ -97.81916

Redtail WMA 41.0897/-96.9997

Renquist Basin WMA 41.02821/-97.69821

Rhoden WMA 41.02023/ -95.86514

Riverview SRA 40.69202/ -95.85046

Rock Creek SRA 40.09322/-101.76296

Rock Creek Station SHP 40.11012/ -97.05809

Rock Creek Station SRA 40.11179/-97.06526

Rock Glen WMA 40.09797/-97.06266

Rockford Lake SRA 40.22091/ -96.58174

Rose Creek WMA 40.07596/ -97.23664

S

Sacramento-Wilcox WMA 40.37246/-99.24107

Saline County Easement WMA 40.39443/ -97.174

Sandpiper WMA 40.49906/-97.71439

Sandy Channel SRA 40.66825/ -99.37529

Schlagel Creek WMA 42.71831/-100.61276

Schramm Park SRA 41.02403/-96.25092

Shady Trail WMA 40.69625/-96.99452

Shell Lake WMA 42.92137/-102.0357

Sherman Reservoir SRA/WMA 41.32469/ -98.90856

Silver Creek WMA 41.32352/-97.62508

Sioux Strip WMA 42.34102/-97.30543

Skull Creek #1 WMA 41.22893/-96.98981

Skull Creek #2 WMA 41.36724/ -96.9559

Smartweed Marsh West WMA 40.34193/ -98.037

Smith Falls SP 42.88822/-100.31526

Smith Lake WMA 42.40563/-102.45414

Sora WMA 40.38186/ -97.65868

South Fork WMA 40.08405/ -95.89999

South Fork WMA 40.08592/ -95.89856
South Pine WMA 42.36858/ -99.72682
South Sacramento WMA 40.31825/ -99.23166
South Twin Lake WMA 42.3137/-100.11954
Southeast Sacramento WMA 40.34412/-99.21895
Spencer Dam WMA 42.80788/ -98.65521
Spikerush WMA 40.91052/ -97.48675
Stagecoach Lake SRA 40.59685/-96.64191
Sunny Hollow WMA 41.37685/ -97.73845
Sunshine Bottoms WMA 42.92093/ -98.40673
Sutherland Reservoir SRA 41.10646/-101.13099
Swan Creek WMA 40.51338/-97.25025
Swanson Reservoir SRA 40.15677/-101.06213
Swanson Reservoir WMA 40.16376/-101.13296

T

Table Rock WMA 40.18366/ -96.06326
Tatanka WMA 42.8175/-97.4701
Taylor's Branch WMA 40.14222/ -96.12577
Teal WMA 40.55646/ -96.8775
Thomas Creek WMA 42.76684/ -99.68877
Thompson-Barnes WMA 42.30442/ -97.04216
Triple Creek WMA 40.66432/ -96.40422
Twin Lakes R.C. WMA 42.31449/-99.47851
Twin Lakes WMA 40.83408/ -96.9534
Twin Oaks WMA 40.33116/-96.14263
Two Rivers SRA 41.21677/-96.35175
Two Rivers WMA 41.20768/ -96.34722

U

Union Pacific SRA 40.67777/-99.25293

V

Verdel Landing WMA 42.82971/-98.15151
Verdon SRA 40.14742/ -95.72422
Victoria Springs SRA 41.60948/ -99.75135

W

Wagon Train SRA 40.63625/-96.58482
Walgren Lake SRA 42.63765/ -102.62868
Wanamaker WMA 40.5413/-101.66187
Wapiti WMA 40.90047/-100.57383
War Axe SRA 40.72391/-98.73473
Wellfleet WMA 40.75867/-100.74806
West Cozad WMA 40.85444/-100.01008
West Elm Creek WMA 40.69411/-99.47828
West Gothenburg WMA 40.97286/ -100.2956
West Hershey WMA 41.13868/-101.01434
West Maxwell WMA 41.05603/ -100.54307
West Sacramento WMA 40.36439/-99.31268
White Front WMA 40.55009/-98.08216
Whitetail WMA 41.40429/-97.08393
Whitney Inlet WMA 42.78603/-103.32354
Wildcat Hills SRA 41.70257/-103.67166
Wildcat Hills WMA 41. 70553/ -103.65875
Wildwood WMA 41.0386/-96.83948
Wilkinson WMA 41.50423/ -97.49384
Willow Creek SRA 42.17531/ -97.57205
Willow Island WMA 40.87635/-100.06319
Willow Lake B.C. WMA 42.23678/ -100.0813
Wilson Creek WMA 40.70654/ -96.05648
Windmill SRA 40.70704/ -98.83886
Wiseman WMA 42.75643/-97.09508
Wood Duck WMA 41.93084/ -97.31538
Wood River West WMA 40.75502/ -98.61092

Y

Yankee Hill WMA 40.7248/ -96.78885
Yellowbanks WMA 42.04354/ -97.66397

Latitude / Longitude Data for Federally Owned Birding Sites

A
Atlanta WPA 40.382/-99.478

B
Bluebill WPA 40.636/-97.703
Bluestem WPA 40,441/-99.058
Brauning WPA 40.6/-97.724

C
Clark WPA 40.378/-99.053
Cottonwood WPA 40.547/-99.584
County Line WPA 40.7027/-97.5419

E
Eckhardt WPA Lat./Long 40.465/-99.903
Elley Lagoon WPA 40.486/-99.684

F
Frerichs WPA 40.372/-99.125
Funk Lagoon WPA 40.498/-99.227

G
Gleason WPA 40.435/-99.025
Glenvil WPA 40.475/-98.22
Green Acres WPA 40.46/-97.938
Greiss WPA 40.584/-97.776

H
Hansen WPA 40.449/-97.848
Harms WPA 49.48972/-98.00972
Harvard WPA 40.614/-98.181
Hultine WPA 40.628/-97.973

J
Jensen WPA 40.4/97.744
Johnson Lagoon WPA 40.556/-99.326
Jones Marsh WPA 40.39/-99.432

K
Kenesaw WPA 40.602/-98.645
Killdeer Basin WPA 40.389/-99.104
Krause WPA 40.472/-97.797
Krause WPA 40.472/-997.797

L
Lange WPA 40.563/-97.846
Lindau WPA 40.402/-99.036

M
Mallard Haven WPA 40.448/-97.744
Massie WPA 40.479/-98.083
Meadowlark WPA 40.4736/-99.995
Millers Pond WPA 40.387/-97.731
Moger WPA 40.483/-978/992
Morphy WPA 40.611/-97.732

N
Nelson WPA 40.759/-97.937
North Hultine (formerly Sandpiper) WPA
    40.499/-97.714

P
Peterson WPA 40.489/-99.658
Prairie Dog WPA 40.402/-99.13

Q
Quadhammer WPA 40.3/-99.1

R
Rauscher WPA 40.585/-97.764
Real WPA 40.672/-97.575
Ritterbush WPA 40.26/-99.043
Rolland WPA 40.59/-97.821

S
Schwisow WPA 40.391/-97.166
Shuck WPA 40.458/-99.997
Sinninger WPA 40.716/-97.536
Smith WPA 40.439/-97.974
Springer W. P. A.. 40.849/-98.128

T
Theesen WPA 40.512/-98.273
Troester Basin WPA 40.797/-97.924

V
Verona WPA 40.512/-98.273
Victor Lake WPA 40.594/-99.654

W
Waco WPA 40.911/-97.479
Weiss WPA 40.455/-97.732
Wilkins WPA 40.609/-97.675

Y
Youngson WPA 40.396/98.785

List of Mapped Birding Sites, by County

Sioux County (Map 1, recently burned areas are outlined)
Fort Robinson State Park (Map location 9)
Gilbert-Baker WMA (Map location 4)
James Ranch SRA (Map location 8)
Nebraska National Forest, Pine Ridge District (Map location 10)
Oglala National Grasslands (Map locations 1, 3)
Peterson WMA (Map location 6)
Soldier Creek Wilderness (Map location 7)
Sowbelly Canyon (Map location 5)
Toadstool Geologic Park (Map location 2)

Dawes County (Map 2, recently burned areas are outlined)
Box Butte SRA (Map location 3)
Chadron State Park (Map location 7)
Fort Robinson State Park (Map locations 5, 8)
Oglala National Grasslands (Map location 1)
Pine Ridge National Recreation Area and Nebraska National Forest, Pine Ridge Unit (Map location 2)
Ponderosa WMA (Map location 6)
Whitney Lake WMA (Map location 4)

Sheridan County (Map 3)
Sandhills marshes near Lakeside (Map location 2).

Scotts Bluff County (Map 4)
Nine Mile Creek Special Use Area (Map location 3)
North Platte National Wildlife Refuge, including Lake Minatare SRA (Map location 2)
Scotts Bluff National Monument (Map location 1)
Wildcat Hills State Recreation Area and Buffalo Creek WMA (Map location 4)

Kimball County (Map 5)
Lodgepole Creek (Map location 3)
Oliver Reservoir SRA (Map location 1)
Tri-state corner (Map location 2)

Morrill County (Map 6)
Bridgeport SRA (Map location 4)
Chimney Rock National Historic Site (Map location 1)
Courthouse Rock and Jail Rock (Map location 6)
Facus Springs WMA (Map location 2)
Redington Gap road (Map location 5) and road south of Redington (Map location 7)
Saline marsh near Bridgeport (Map location 3)

Garden County (Maps 3, 7)
Ash Hollow State Historical Park (Map 7, locations 1 & 3)
Clear Creek Waterfowl Management Area (Map 7, location 2)
Crescent Lake National Wildlife Refuge (Map 3, location 1)

Deuel County (Map 7)
Bittersweet WMA (Map location 6)
Goldeneye WMA (Map location 5)
Goldenrod WMA (Map location 4)
Merritt Reservoir WMA (Map location 4)
Samuel R. McKelvie District, Nebraska National Forest (Map location 2)
Schlegel Creek WMA (Map location 3)
Valentine National Wildlife Refuge (Map location 7)

Keya Paha County (Map 9)
Cub Creek Recreation Area (Map location 2)
Thomas Creek WMA (Map location 3)

Brown County (Map 10)
American Game Marsh WMA (Map location 8)
Bobcat WMA (Map location 3)
Long Lake WMA (Map location 9)
Long Pine WMA (Map location 6)
Niobrara Valley Preserve (Map location 1; headquarters at location 2)

Pine Glen WMA (Map location 5)
School Land WPA & Keller Park SRA (Map location 4)
South Twin Lake WMA. (Map location 7)
Willow Lake WMA (Map location 10)

Thomas County (Map 11)
Bessey Division, Nebraska National Forest (Map location 1; locations 2 and 3 indicate lek sites)

Blaine County (Map 11)
Bessey Division, Nebraska National Forest (Map location 1)

Arthur County (Map 12)
Marshes near McPherson County border (Map location 1)

Keith County (Map 13)
Cedar Point Biological Station (Map location 4)
Clear Creek WMA (Map location 1)
Kingsley Dam and Lake Ogallala SRA (Map location 3)
Lake McConaughy SRA (Map location 2)
Lakeview SRA Area (Map location 6)
Ogallala Strip WMA (Map location 5)

Lincoln County (Map 14)
Birdwood Lake WMA (Map location 9)
Chester Island WMA (Map location 14)
East Hershey WMA (Map location 8)
East Sutherland WMA (Map location 6)
Fremont Slough WMA (Map location 10)
Ft. McPherson Cemetery (Map location 12)
Hershey WMA (Map location 7)
Jeffrey Canyon WMA and Reservoir (Map location 3)
Malony Reservoir SRA (Map location 2)
Muskrat Run WMA (Map location 5)
North Platte Sewage lagoons (Map location 15)
North River Wildlife WMA (Map location 4)
Platte WMA (Map location 11)
Sutherland Reservoir SRA (Map location 1)
West Brady WMA (Map location 13)

Dawson County (Map 15)
Cozad WMA (Map location 4)
Darr Strip WMA (Map location 6)
Dogwood WMA (Map location 7)
East Cozad WMA (Map location 5)
East Willow Island WMA (Map location 2)
Gallagher Canyon SRA (Map location 9)

Johnson Lake SRA (Map location 11)
Midway Lake WMA (Map location 8)
Plum Creek WMA (Map location 10)
West Cozad WMA (Map location 3)
Willow Island WMA (Map location 1)

Chase County (Map 16)
Champion Lake SRA (Map location 4)
Enders Reservoir SRA (Map location 1)
Enders Reservoir WMA (Map location 2)
Wannamaker WMA (Map location 3)

Frontier County (Map 17)
Medicine Creek Reservoir & Medicine Creek SRA/ WMA (Map location 2)
Red Willow Reservoir SRA/WMA (Map location 1)

Gosper County (Map 18)
Elley Lagoon Federal Waterfowl Area (Map location 6)
Johnson Lake SRA (Map location 1, main lake not shown)
Peterson Basin Federal Waterfowl Area (Map location 7)
Victor Lake Federal Waterfowl Area (Map location 2)

Phelps County (Map 18)
Atlanta Marsh WPA (Map location 9)
Cottonwood Basin WPA (Map location 3)
Funk Lagoon WPA (Map location 8)
Johnson Lagoon WPA (Map location 5)
Jones Marsh WPA (Map location 10)
Linder WPA (Map location 4)
Sacramento-Wilcox WMA (Map location 12)
West Sacramento WMA (Map location 11)

Hitchcock County (Map 19)
Swanson Reservoir WMA (Map location 1)

Harlan County (Map 20)
Harlan County Dam (Map location 1)
South Sacramento Wildlife Area (Map location 2)
Southeast Sacramento Wildlife Area (Map location 3)

Knox County (Map 21)
Bazille Creek WMA (Map location 2)
Bohemia Prairie WMA (Map location 4)
Gavin's Point Dam, Lewis & Clark Lake SRA (Map location 1)
Niobrara State Park (Map location 3)

Antelope County (Map 22)
Ashfall Fossil Beds State Historical Park (Map location 1)
Grove Lake WMA (Map location 2)
Hackberry Creek Public Use Area (Map location 3)
Redwing WMA (Map location 4)

Pierce County (Map 23)
Willow Creek SRA (Map location 1)

Madison County (Map 23)
Oak Valley WMA (Map location 3)
Yellowbanks WMA (Map location 2)

Platte County (Map 25)
George Syas WMA (Map location 4)
Lake Babcock Waterfowl Refuge and Lake Babcock (Map location 6)
Lake North (Map location 7)

Sherman County (Map 24)
Bowman Lake SRA (Map location 2)
Sherman Reservoir SRA/WMA (Map location 1)

Nance County (Map 25)
Loup Lands WMA (Map locations 1, 2)
Prairie Wolf WMA (Map location 3)
Sunny Hollow WMA (Map location 5)

Buffalo County (Map 26)
Bassway Strip WMA (Map location 10)
Blue Hole WMA (Map location 2)
Cottonmill Lake Public Use Area (Map location 6)
East Odessa SRA (Map location 5)
Gibbon Bridge (Map location 11)
Lillian Annette Rowe Sanctuary & Iain Nicolson Audubon Center (Map location 7)
Ravenna Lake SRA (Map location 1)
Sandy Channel SRA (Map location 3)
Union Pacific SRA (Map location 4)
War Axe SRA (Map location 8)
Windmill SRA (Map location 9)

Hall County (Map 27)
Amick Acres Road (Map location 8)
Cornhusker WMA (Map location 2)
Hall County Park & Stuhr Museum (Map location 12)
Mormon Island SRA (Map location 10)
Nine-mile Bridge (Map location 9)
Shoemaker Island Road. (Map location 4)
Taylor Ranch Road (Map location 1)

Hamilton County (Map 28)
Deep Well WMA (Map location 2)
Gadwall WMA (Map location 1)
Nelson WPA (Map location 5)
Pintail WMA (Map location 4 left)
Springer WPA (Map location 3)
Troesler Basin WPA (Map location 4 right)

Clay County (Map 28)
Eckhart Lagoon WPA (Map location 20)
Glenvil Basin WPA (Map location 13)
Green Acres WPA (Map location 19)
Greenhead WPA (Map location 22)
Greenwing WPA (Map location 24)
Hansen Lagoon WPA (Map location 23)
Harms WPA (Map location 16)
Harvard Marsh WPA (Map location 8)
Hultine WPA (Map location 7)
Kissinger Basin WPA (Map location 14)
Lange Lagoon WPA (Map location 10)
Massie Lagoon WPA (Map location 12)
McMurtry Refuge (Map location 9)
Meadowlark WPA (Map location 15)
Moger WPA (Map location 17)
North Hultine (formerly Sandpiper) WPA (Map location 6)
Shuck WPA (Map location 18)
Smith Lagoon WPA (Map location 21)
Theesen Lagoon WPA (Map location 11)

Fillmore County (Map 29)
Bluebill WPA (Map location 10 right) & Marsh Hawk WPA (Map location 10 left)
County Line Marsh WPA (Map location 8)
Krause Lagoon WPA (Map location 16)
Mallard Haven WPA (Map location 17)
Murphy Lagoon WPA (Map location 12)
Rauscher Lagoon WPA (Map location 14)
Real WPA (Map location 9)
Rolland Lagoon WPA (Map location 13)
Sandpiper WMA (Map location 15)
Weiss Lagoon WPA (Map location 18)
Wilkins Lagoon WPA (Map location 11)

York County (Map 29)
County Line Marsh WPA (Map location 8)
Kirkpatrick Basin WMAs (Map location 5 & 6)
Sinninger Lagoon WPA (Map location 7)
Spikerush WMA (Map location 4)
Waco Basin WPA (Map location 3)

Franklin County (Map 30)
Macon Lakes WPA (Map location 15)
Quadhammer Marsh WPA (Map location 14)
Ritterbush Marsh WPA (Map location 13)

Kearney County (Map 30)
Bluestem Basin WPA (Map location 4)
Clark Lagoon WPA (Map location 12)
Fort Kearny SRA (Map location 2)
Frerichs Lagoon WPA (Map location 10)
Gleason Lagoon WPA (Map location 5)
Hike-Bike bridge (Map location 1)
Jensen Lagoon WPA (Map location 9)
Killdeer Basin WPA (Map location 11)
Lindau Lagoon WPA (Map location 7)
Northeast Sacramento WMA (Map location 7)
Prairie Dog Marsh WPA (Map location 6)
Youngson Lagoon WPA (Map location 8)

Adams County (Map 31)
Ayr Lake (Map location 5)
Crystal Lake SRA (Map location 4)
DLD SRA (Map location 6)
Hastings Museum and Lake Hastings (Map location 7)
Kenesaw Lagoon (Map location 1)
Little Blue River (Map location 2)
Prairie Lake Public Use Area (Map location 3)

Dixon County (Map 32)
Ponca State Park (Map location 1)

Dakota County (Map 32)
Basswood Ridge WMA (Map location 2)
Omadi Bend WMA (Map location 3)

Dodge County (Map 33)
Dead Timber SRA (Map location 1, right)
Fremont Lakes SRA (Map location 2)
Powder Horn WMA (Map location 1, left)

Washington County (Map 34)
Boyer Chute National Wildlife Refuge (Map location 3)
DeSoto National Wildlife Refuge (Map location 1)
Fort Atkinson State Historic Park (Map location 2)

Saunders County (Map 35)
Jack Sinn WMA (Map location 7)
Larkspur WMA (Map location 3)
Pahuk Natural Area (Map location 1)

Red Cedar Public Use Area (Map location 2)

Douglas County (Map 36)
Glenn Cunningham Lake (Map location 6)
Neale Woods Nature Center (Map location 7)
Standing Bear Lake (Map location 5)
Two Rivers SRA (Map location 1)
Zorinsky Lake (Map location 9)

Sarpy County (Map 36)
Chalco Hills Recreation Area (Map location 2)
Fontenelle Forest Preserve (Map location 4)
Gifford Point (Map location 8)
Schramm Park Recreation Area (Map location 3)

Seward County (Maps 29, 37)
Branched Oak SRA (Map 37, location 3)
Bur Oak WMA (Map 37, location 4)
Freeman Lakes WMA (Map 29, location 1)
Meadowlark NRD Recreation Area (Map 37, location 1)
North Lake Basin WMA (Map 29, location 2)
Oak Glen WMA (Map 37, location 2)
Twin Lakes WMA (Map 37, location 5)

Lancaster County (Maps 35, 38)
Arbor Lake (Map 38, location 1)
Bluestem Lake SRA (Map 38, location 11)
Roper's Lake (undeveloped; Map 38, location 2)
Branched Oak Lake SRA (Map 35, location 5)
Capital Beach Saline Wetlands (Map 38, location 6)
Conastoga Lake SRA (Map 38, location 8)
Cottontail Public Use Area (Map 38, location 14)
Hedgefield Lake WMA (Map 38, location 18)
Holmes Lake (Map 38, location 16)
Jack Sinn WMA (Map 35, locations 6 & 7)
Killdeer WMA (Map 38, location 10)
Nine-Mile Prairie (Map 38, location 3)
Oak Lake Park (Map 38, location 5)
Olive Creek SRA (Map 38, location 12)
Pawnee Lake SRA (Map 38, location 4)
Pioneer's Park (Map 38, location 7)
Spring Creek Prairie Audubon Center (Map 38, location 19).
Stagecoach Lake SRA (Map 38, location 15)
Teal WMA (Map 38, location 13)
Wagontrain Lake SRA (Map 38, location 17)
Wilderness Park (Map 38, location 14)
Wildwood Lake WMA (Map 35, location 4)
Yankee Hill Lake SRA (Map 38, location 9)

Cass County (Map 39)
Eugene T. Mahoney State Park (Map location 1)
Louisville Lakes State Park (Map location 3)
Platte River State Park (Map location 2)

Gage County (Map 40)
Arrowhead WMA (Map location 6)
Big Indian Public Use Area (Map location 8)
Claytonia Public Use Area (Map location 1)
Diamond Lake WMA (Map location 7)
Homestead National Monument of America (Map
    location 2)
Iron Horse Trail (Map location 3)
Rockford Lake SRA (Map location 4)
Wolf Wildcat Public Use Area (Map location 5)

Johnson County (Map 41)
Hickory Ridge WMA (Map location 4)
Osage WMA (Map locations 1-3)
Twin Oaks WMA (Map locations 5-7)

Pawnee County (Map 42)
Bowwood WMA (Map location 3)
Burchard Lake WMA (Map location 2)
Iron Horse Trail (Map location 1)
Pawnee Prairie WMA (Map location 4)
Prairie Knoll WMA (Map location 5)

Nemaha County (Map 43)
Indian Cave State Park (Map location 1)

Richardson County (Map 43)
Four Mile Creek WMA (Map location 4)
Indian Cave State Park (Map location 1)
Iron Horse Trail (Map location 5)
Kinters Ford WMA (Map location 3)
Rulo Bluffs Preserve (Map location 7)
Verdon Lake SRA (Map location 2)

Otoe County (Map 44)
Arbor Lodge SHA (Map location 2)
Missouri River Basin Lewis and Clark Interpretive
    Trail and Visitor Center (Map location 3)
Triple Creek WMA (Map location 4)
Wilson Creek WMA (Map location 1)

Species Guide to Refuges, Sanctuaries and Other Birding Sites

The following pages provide a tabular guide to the typical locations of 348 species associated with ten major birding areas in Nebraska. The ten areas include (left half) five national wildlife refuges or other federally managed areas, organized from west to east. These bird listings are based on refuge checklists. The right half includes mostly non-federal locations, also organized from west to east. The final column is available as a personal checklist. The bird lists are based on various sources, as indicated below. They are abbreviated as follows:

Federally Managed Areas

C.L.R. = Crescent Lake National Wildlife Refuge, Garden Co.

Val. R. = Valentine National Wildlife Refuge, Cherry Co.

F.N.R. = Fort Niobrara National Wildlife Refuge, Cherry Co.

R.W.B. = Rainwater Basin Wetland Management District, southeastern Nebraska

D.S.R. = De Soto National Wildlife Refuge (partly in Iowa)

Non-Federal Areas

NW Ne. = Northwestern Nebraska (Sioux, Dawes, Box Butte & Sheridan Counties), based on Rosche, 1982

N.P.V. = North Platte Valley, from Scottsbluff to North Platte, including North Platte National Wildlife Refuge and especially the environs of Lake McConaughy (based on refuge leaflets, Rosche, 1994, and Brown & Brown, 2000)

P.R.V. = Platte River Valley, from North Platte to Grand Island (based on Johnsgard, 1984, Lingle, 1994, and personal observations)

Lanc. = Lancaster County, including bird lists for Pioneer's Park, Wilderness Park, Lancaster County, and personal observations.

Omaha = Omaha vicinity, based on bird lists for Fontenelle Forest and Neale Woods, plus observations of Babs & Loren Padelford (personal communication).

_____ = An additional column for use as a personal checklist.

The abbreviations given for each species provide estimates of its overall abundance and seasonal status. The first letter indicates relative abundance, as follows:

A = Abundant

C = Common

U = Uncommon

O = Occasional

R = Rare

V = Vagrant (indicating species well out of their normal ranges)

? = Indicates a questionable attribution of status for the area.

Accidental records, based on a single sighting at the indicated locality, are excluded.

The remaining letter or letters indicate seasonal status as follows:

M = Migrant during spring and fall (relative abundance estimate are for the the season of greater abundance, when these differ).

SpM = Spring Migrant, with few or no fall records.

FM = Fall Migrant, with few or no spring records.

WM = Wintering Migrant, usually arriving during fall and remaining until spring.

PR = Permanent Resident, typically remaining all year. Abundance estimates for such species relate primarily to the summer period.

SR = Summer Resident, present during the breeding season and potentially but not necessarily breeding locally. Relative abundance estimates relate to the summer period; abundance during spring and fall migration periods may differ considerably.

* = Breeding has been recorded for the location.

114

Refuge, Region, or Area	Federal Refuges or Managed Areas					Non-Federal Lands or Regions				
	C.L.R.	VAL.R.	F.N.R.	R.W.B.	D.S.R.	NW.Ne	N.P.V.	P.R.V.	LANC.	OMAHA
DUCKS, GEESE, SWANS										
Tundra Swan	OM	RM	--	OM	RM	RSpM	RFM	--	RM	RM
Trumpeter Swan	RPR*	RPR*	--	OM	RM	RSR*	RM	--	RM	--
Greater White-fronted Goose	OM	OM	--	AM	UM	RM	RM	AM	CM	UM
Ross' Goose	--	--	--	--	UM	UM	OM	UM	OM	--
Snow Goose	UM	OM	--	CM	AM	CM	UM	AM	AM	CM
Cackling Goose	UM	UM	CM	CM	CM	CM	CM	CM	CM	CM
Canada Goose	CPR*	UPR*	CM	AM	CPR*	UPR	CPR*	CPR*	CPR*	CM
Wood Duck	RSR	RSR	RSR	USR*	CSR*	OSR	USR*	USR*	CSR*	CSR*
Green-winged Teal	USR*	USR	USR	USR*	CSR*	USR*	UWM	USR	CM	CM
American Black Duck	--	OM	--	--	UM	RSpM	--	--	RM	--
Mallard	CPR*	APR*	CPR*	CPR*	CPR*	CPR*	CPR*	CPR*	CPR*	CSR
Northern Pintail	CSR*	CSR*	USR*	CPR*	CM	CSR*	UPR*	USR	OSR*	OM
Blue-winged Teal	ASR*	ASR*	CSR*	CSR*	USR*	CSR*	OSR*	CSR*	OSR	USR
Cinnamon Teal	RSR*	RSR	RM	OSR*	--	USpM	OSM	--	RM	OM
Northern Shoveler	CSR*	CSR*	CSR*	CSR*	CM	CSR*	UM	RSR*	CM	CM
Gadwall	CSR*	CSR*	CSR*	CSR*	CM	CSR*	UM	USR*	CM	CM
Eurasian Wigeon	--	RM	--	--	--	--	--	--	--	--
American Wigeon	USR*	USR*	OSR*	OSR*	CM	USR*	UM	CM	CM	CM
Canvasback	USR*	OSR*	OSR	OSR*	CM	USR	UM	UM	UM	OM
Redhead	USR*	USR*	USR*	USR*	UM	USR*	AM	CM	CM	RM
Ring-necked Duck	OSR	RSR	OSR	UM	CM	CM	UM	RSpM	CM	CM
Greater Scaup	OSpM	RM	--	RSpM	OM	--	RWM	--	RM	--
Lesser Scaup	USR*	CM	CM	AM	CM	CM	UM	USpM	CM	UM
Oldsquaw	--	--	--	--	--	--	RWM	--	RM	--
Black Scoter	--	--	--	RM	--	--	RFM	--	--	--
Surf Scoter	--	--	--	--	--	RFM	RWM	--	RM	--
White-winged Scoter	--	--	--	--	RM	RFM	RWM	--	RM	--
Barrow's Goldeneye	--	--	--	--	--	--	RWM	--	--	--
Common Goldeneye	UM	OM	UM	UM	UM	CM	CWM	UWM	UM	CM
Bufflehead	CM	CM	UM	UM	UM	CM	CWM	UM	UM	UM
Hooded Merganser	RM	OM	OM	UM	UM	UM	RWM	RFM	OM	--
Common Merganser	UM	CM	CM	OM	CM	CWM	CWM	CWM	CWM	CWM
Red-breasted Merganser	RM	RM	OM	--	OM	RM	RWM	--	RM	OM
Ruddy Duck	CSR*	CSR*	UM	USR*	UM	USR*	UM	--	CM	OM
GALLIFORM BIRDS										
Gray Partridge	--	RPR	--	--	OWV	--	--	--	--	--
Ring-necked Pheasant	CPR*	APR*	CPR*	CPR*	APR*	UPR*	CPR*	CPR*	CPR*	UPR
Sage Grouse	--	--	--	--	--	RPR	--	--	--	--
Greater Prairie-chicken	--	UPR*	UPR*	--	--	--	UPR*	UPR*	--	--
Sharp-tailed Grouse	CPR*	APR*	CPR*	--	RV	CPR*	CPR*	CPR*	--	--
Wild Turkey	--	RPR	CPR*	--	CPR*	CPR*	UPR*	CPR*	CPR*	UPR
Northern Bobwhite	OPR*	RPR*	UPR*	CPR*	CPR*	RPR	UPR*	CPR*	CPR*	OPR
LOONS										
Common Loon	OM	RM	--	OM	OM	RSpM	UM	--	OM	RM
Red-throated Loon	--	--	--	--	--	--	--	--	RM	--
GREBES										
Pied-billed Grebe CSR*	CSR*	RSR	CSR*	USR	CSR*	USR*	USR*	USR	CM	
Horned Grebe	UM	RM	UM	UM	OM	USpM	UM	--	RM	--

Refuge, Region, or Area	Federal Refuges or Managed Areas					Non-Federal Lands or Regions				
	C.L.R.	VAL.R.	F.N.R.	R.W.B.	D.S.R.	NW.Ne	N.P.V.	P.R.V.	LANC.	OMAHA
Eared Grebe	CSR*	CSR*	USR	OSR	OM	CSR*	CM	UM	UM	UM
Western Grebe	CSR*	CSR*	USR	--	OM	CSR	CSR*	--	RM	--
Clark's Grebe	OSR	--	--	--	--	--	OSR*	--	--	--
PELICANS & CORMORANTS										
American White Pelican	USR	CSR	USR	OSR	RSR	CSR	CSR	UM	CM	CM
Double-crested Cormorant	CSR*	ASR*	OSR	CM	CM	USR*	CSR*	UM	CM	CM
HERONS										
American Bittern	CSR*	CSR*	USR	CSR*	USR	USR*	USR*	USR*	RSR	UM
Least Bittern	RSR*	RSR	--	RSR	OSR	--	RSR	--	RSR*	OM
Great Blue Heron	USR*	CSR*	CSR	CSR*	USR	CSR*	CSR*	CSR*	USR	CSR
Great Egret	OSR	OSR	--	OM	USR	RSR	OM	OSR	RM	UM
Snowy Egret	OSR	OSR	--	OM	UM	RSpV	RM	--	RM	UM
Little Blue Heron	--	RSR	--	RM	RSpM	--	RM	OM	RM	UM
Cattle Egret	OSR	RSR	--	OSR	RM	RSpV	USR*	--	OM	UM
Green Heron	OSR	RSR	--	USR*	USR	USR	USR	USR	USR*	CSR
Black-crowned Night Heron	CSR*	ASR*	CSR	OSR	RSR	RSR*	OM	USR*	UM	USR
Yellow-crowned Night Heron	--	--	--	RM	RSR	--	--	--	RM	UM
IBISES										
White-faced Ibis	OSR	OSR	--	RM	--	UM	OM	--	--	--
VULTURES										
Turkey Vulture	OSR	OSR	CSR	UCR	USR	CSR*	CSR*	USR*	CSR*	CSR
HAWKS, EAGLES, KITES										
Osprey	OM	RM	RM	OM	UM	RM	UM	OM	UM	UM
Mississippi Kite	--	--	--	--	--	--	OSR*	--	OM	RM
Bald Eagle	RM	OM	UM	UM	CM	UWM	CWM	CWM	UWM	UWM
Northern Harrier	UPR*	CPR*	CSR	CPR*	UPR	UPR*	UPR*	OPR*	UPR	UPR
Sharp-shinned Hawk	OPR	OM	UM	UPR	UM	UWM	OWM	UM	UPR	UM
Cooper's Hawk	OSR	RPR	OPR	UPR	UPR	RPR	RM	UM	RPR	UM
Northern Goshawk	RM	RWM	OWM	OM	RM	RM	RWM	OM	RWM	RWM
Red-shouldered Hawk	--	RM	--	--	RM	RSpM	--	--	RM	UM
Broad-winged Hawk	RM	RM	--	--	RM	RSpM	RSpM	--	OM	UM
Swainson's Hawk	USR*	USR*	USR	USR*	OM	USR*	CSR*	USR*	UM	UM
Red-tailed Hawk	UPR*	UPR*	CSR*	CPR*	CPR*	CPR*	CSR*	UPR*	CPR*	CR*
Ferruginous Hawk	OSR	OPR*	--	UM	--	OPR*	OM	OM	RM	--
Rough-legged Hawk	UM	CWM	UM	UM	--	UWM	UWM	UWM	RWM	UWM
Golden Eagle	RPR	OM	UM	UM	OM	UPR*	RPR*	UWM	RWM	RM
FALCONS										
American Kestrel	RPR	UPR	CPR	CPR*	UPR*	CPR*	UPR*	CPR	CPR*	UPR
Merlin	RM	OM	RM	UM	--	UWM	OWM	OM	RM	OM
Peregrine Falcon	--	--	--	RM	RM	--	RM	RM	RM	OSR*
Gyrfalcon	--	--	--	RWM	--	--	--	--	--	--
Prairie Falcon	RPR	RM	UM	OM	OM	UPR*	RPR	RPR	RM	RWM

Refuge, Region, or Area	Federal Refuges or Managed Areas					Non-Federal Lands or Regions				
	C.L.R.	VAL.R.	F.N.R.	R.W.B.	D.S.R.	NW.Ne	N.P.V.	P.R.V.	LANC.	OMAHA
RAILS, COOTS							CPR*			
Yellow Rail	--	--	--	RSpM	--	--	--	--	RFM	--
Black Rail	--	--	--	RSpM	--	--	--	--	RFM	--
King Rail	--	--	--	RSpM	--	--	RSR	--	RSR	--
Virginia Rail	CSR*	CSR*	RSR*	USR*	RSR	USR*	USR*	OSR*	RSR	USR
Sora	CSR*	CSR*	RSR	CSR*	OSR	USR*	USR*	USR*	USR*	UM
Common Moorhen	--	RSR	--	--	--	--	--	--	RSR	RM
American Coot	CSR*	ASR*	USR	ASR*	USR	CSR*	USR*	CSR*	UPR*	CSR
CRANES										
Sandhill Crane	RM	CM	UM	AM	RM	AM	AM	AM	RSpM	--
Whooping Crane	--	RM	--	OM	--	--	RM	RM	--	--
PLOVERS										
Black-bellied Plover	RM	RM	--	OM	RM	UFM	OM	OM	UM	--
American Golden-plover	OM	--	--	OM	RSpM	RFM	--	OM	RM	--
Snowy Plover	--	--	--	RSR	--	RSM	RSR	RSR	--	--
Semipalmated Plover	OM	RM	OM	CM	RM	CM	UM	UM	CM	OM
Piping Plover	--	OSR	OM	CM	RM	RSpM	USR*	CSR*	UM	--
Killdeer	CPR*	CSR*	CSR*	CPR*	CPR*	CSR*	CSR*	CPR*	CPR*	CSR
AVOCETS, STILTS										
Black-necked Stilt	OSR	--	--	--	--	--	RM	RM	--	--
American Avocet	CSR*	CSR*	RSR	USR*	RSR	CSR*	UM	USR*	RM	--
SANDPIPERS, SNIPES										
Greater Yellowlegs	UM	UM	UM	CM	CM	CM	CM	CM	UM	UM
Lesser Yellowlegs	CM	CM	CM	CM	CM	CM	CM	CM	CM	UM
Solitary Sandpiper	UM	OM	OM	CM	OM	UM	CM	UM	UM	CM
Willet	USR*	USR*	OSR	CM	UM	CSR*	UM	UM	RM	--
Spotted Sandpiper	CSR*	CSR*	CSR*	USR*	CSR*	CSR*	USR*	CSR*	USR	CM
Upland Sandpiper	CSR*	CSR*	CSR*	USR*	OSR	CSR*	CSR*	CSR*	RSR*	RM
Whimbrel	--	--	--	RSpM	--	RM	RSpM	RM	RSpM	--
Long-billed Curlew	CSR*	USR	CSR*	USR*	--	CSR*	USR*	RM	--	--
Hudsonian Godwit	--	--	--	CSpM	OSpM	--	RSpM	OSpM	RM	--
Marbled Godwit	RM	OM	UM	UM	RM	UM	UM	OM	RM	--
Ruddy Turnstone	--	--	--	OSpM	RSpM	RFM	RSpM	RM	RM	--
Red Knot	RM	RM	--	--	RSpM	RFM	RM	RSpM	RFM	--
Sanderling	RM	RM	--	--	RSpM	UFM	OM	OM	--	--
Semipalmated Sandpiper	UM	UM	--	CM	OSpM	UM	CM	CM	UM	UM
Western Sandpiper	OM	UM	UM	RM	RSpM	UM	CM	OM	UM	OM
Least Sandpiper	UM	CM	UM	CM	UM	CM	CM	CM	CM	UM
White-rumped Sandpiper	OSpM	UM	OM	CM	UM	USpM	CSpM	CM	UM	OM
Baird's Sandpiper	CM	CM	UM	CM	UM	CM	AM	CM	CM	UM
Pectoral Sandpiper	UM	UM	OM	AM	--	CFM	UM	CM	CM	UM
Dunlin	OFM	RM	RM	OM	RSpM	--	RSpM	OSpM	OM	--
Stilt Sandpiper	UM	RM	--	CM	RSPM	CM	UM	OM	UM	OM
Buff-breasted Sandpiper	--	--	--	RSpM	--	RFM	RFM	OM	RM	--
Short-billed Dowitcher	--	--	--	CM	RM	RM	UM	RM	UM	--
Long-billed Dowitcher	CM	--	UM	UM	OM	CM	CM	UM	CM	--
Wilson's Snipe	RSR*	USR	USR	CM	UM	UPR	UPR	CPR*	UPR	UM
American Woodcock	--	--	--	OM	USR*	--	OM	CSR*	OSR*	USR
Wilson's Phalarope	CSR*	CSR*	UM	CM	OM	CSR*	CM	CSR*	--	UM

	Federal Refuges or Managed Areas					Non-Federal Lands or Regions				
Refuge, Region, or Area	C.L.R.	VAL.R.	F.N.R.	R.W.B.	D.S.R.	NW.Ne	N.P.V.	P.R.V.	LANC.	OMAHA
Red-necked Phalarope	UM	RM	--	--	RSpM	CM	OM	OM	--	--
GULLS, TERNS										
Franklin's Gull	CM	CM	UM	AM	UM	CM	CM	CM	CM	CM
Bonaparte's Gull	RM	RM	--	RM	OFM	RM	UM	OSpM	UM	OM
Ring-billed Gull	UPR	CM	CSR	UM	UPR	CSR	CPR	CPR	UPR	CM
Glaucous Gull	--	--	--	--	--	--	RWM	--	RSpM	--
California Gull	--	--	--	--	--	RM	UPR	--	--	--
Thayer's Gull	--	--	--	--	--	--	RWM	--	--	--
Herring Gull	RM	OM	--	UM	UM	RM	UPR	RM	OM	RM
Caspian Tern	--	--	--	--	OM	--	USR	RSR	RM	--
Common Tern	OSR	CSR*?	USR	UM	RSpM	RM	RM	RSpM	UM	--
Forster's Tern	CSR*	CSR	USR	USR*	USR	USR*	USR	CSR	CM	RM
Least Tern	--	RSR	--	RM	OSR*	--	USR*	CSR*	RM	--
Black Tern	CSR*	ASR*	USR	CSR*	CSR	CSR*	CM	CSR	CSR	UM
PIGEONS, DOVES										
Rock Pigeon	--	--	--	CPR*	--	CPR*	CPR*	CPR*	CPR*	CPR
Mourning Dove	ASR*	CSR*	ASR*	APR*	APR*	CSR*	CPR*	CPR*	CPR*	CPR
CUCKOOS										
Black-billed Cuckoo	OSR	OSR*	USR*	CSR*	USR*	OSR	USR*	USR*	RSR*	UCR
Yellow-billed Cuckoo	USR*	OSR*	OSR	CSR*	CSR*	USR	CSR*	USR*	USR*	CSR
OWLS										
Barn Owl	RSR*	RM	--	OPR	--	RSR*	UPR*	OPR*	RPR*	--
Eastern Screech-owl	OSR	UPR*	UPR*	UPR*	UPR*	UPR*	UPR	CPR*	UPR*	UPR*
Great Horned Owl	UPR*	CPR*	CPR*	CPR*	CPR*	UPR*	CPR*	CPR*	UPR*	UPR*
Snowy Owl	OWM	RWM	RWM	OWM	RWM	RWM	RWM	RWM	RWM	--
Burrowing Owl	USR*	OSR*	CSR*	USR*	--	USR*	USR*	USR*	RSR*	--
Barred Owl	--	--	--	--	--	RV	--	RSR	UPR*	CPR*
Long-eared Owl	OWM	RSR	UPR	UPR*	RPR*	RPR*	OPR*	RFM	RPR*	RPR
Short-eared Owl	UPR*	CPR	UPR	UPR*	--	RSR	UM	UM	RM	--
Northern Saw-whet Owl	--	OSR*?	--	--	--	RSR	--	RSpM	RM	RM
NIGHTJARS										
Common Nighthawk	CSR*	CSR*	CSR*	CSR*	USR	USR*	CSR*	CSR*	CSR*	USR
Common Poorwill	--	--	--	--	--	USR*	USR	RFM	RM	--
Chuck-will's Widow	--	--	--	--	--	--	--	RSR?	--	--
Whip-poor-will	OSR	--	USR*	--	USR*	--	--	--	--	USR*
SWIFTS										
Chimney Swift	OSR	--	--	ASR*	USR*	USR	CSR	CSR*	CSR*	CSR
White-throated Swift	--	--	--	--	--	USR*	--	--	--	--
HUMMINGBIRDS										
Ruby-throated Hummingbird	--	RSR	OSR	USR	USR	--	--	OSR	RSR	USR*
Broad-tailed Hummingbird	--	--	--	--	--	RFM	--	--	--	--
Rufous Hummingbird	--	--	--	--	--	RFM	--	--	--	--

	Federal Refuges or Managed Areas					Non-Federal Lands or Regions				
Refuge, Region, or Area	C.L.R.	VAL.R.	F.N.R.	R.W.B.	D.S.R.	NW.Ne	N.P.V.	P.R.V.	LANC.	OMAHA
KINGFISHERS										
Belted Kingfisher	RSR	OSR*	USR*	CPR*	UPR*	UPR*	UPR*	CPR*	UPR*	CPR
WOODPECKERS										
Lewis' Woodpecker	--	--	--	--	--	RSR*	--	--	--	--
Red-headed Woodpecker	OSR*	OSR*	CSR*	CPR*	APR*	USR*	USR*	CPR*	CPR*	CPR*
Red-bellied Woodpecker	--	--	--	CPR*	CPR*	RV	RPR*	UPR*	UPR*	CPR*
Yellow-bellied Sapsucker	RM	--	--	--	--	--	UM	OM	UWM	UWR
Downy Woodpecker	UPR*	UPR	CPR*	CPR*	CPR*	UPR*	CPR*	CPR*	CPR*	CPR*
Hairy Woodpecker	OPR	OPR	CPR*	CPR*	CPR*	UPR*	UPR*	UPR*	UPR*	CPR*
Northern Flicker	UPR*	CSR*	CSR*	CPR*	CPR*	UPR*	CPR*	CPR*	CPR*	CPR*
Pileated Woodpecker	--	--	--	--	--	--	--	--	--	RWM
TYRANT FLYCATCHERS										
Olive-sided Flycatcher	OM	--	--	UM	--	RSM	RM	RSpM	RM	UM
Western Wood-pewee	USR	--	CSR*	CSR*	--	RSR*	RSR*	--	--	--
Eastern Wood-pewee	OSR	--	--	CSR*	USR*	RSR	RSR*	USR*	USR*	CSR*
Yellow-bellied Flycatcher	--	OM	--	USM	OSR*	--	RSpM	RSpM	RM	UM
Acadian Flycatcher	--	--	USR	--	OSR	--	RSPM	--	UM	OM
Alder Flycatcher	--	--	--	--	OSR	--	CM?	--	RM	UM
Willow Flycatcher	RSR	--	--	--	OSR*	RSR	CSR*	CSR*	RSR*	USR
Least Flycatcher	RSR	RM	--	RSR	CSM	RSR	CM	UM	UM	USR
Cordilleran Flycatcher	RM	--	--	--	--	RSR*	--	--	--	--
Eastern Phoebe	OSR	--	OSR*	CSR*	USR*	USR*	CSR*	USR*	USR*	CSR
Say's Phoebe	OSR	OSR	CSR*	CSR*	--	CSR*	USR*	OSR*	--	--
Great Crested Flycatcher	OSR	RSR*	USR*	OSR*	USR*	RSR*	USR*	CSR*	USR*	CSR*
Cassin's Kingbird	--	--	--	--	--	RSR	RSR*	--	--	--
Western Kingbird	CSR*	CSR*	CSR*	ASR*	ASR*	CSR*	CSR*	CSR*	CSR*	USR
Eastern Kingbird	CSR*	CSR*	CSR*	ASR*	ASR*	CSR*	CSR*	CSR*	CSR*	CSR
Scissor-tailed Flycatcher	--	--	--	RSR*	--	--	RSV	--	OSR	--
LARKS										
Horned Lark	APR*	CPR*	APR*	CPR*	UPR*	CPR*	CPR*	CPR*	UPR*	--
SWALLOWS										
Purple Martin	OSR	--	--	CSR*	CSR*	RSV	USR*	CSR*	CSR*	USR*
Tree Swallow	OSR*	OSR*	CSR	OSR*	USR*	RSR*	USR*	CSR*	CSR*	CSR*
Violet-green Swallow	--	--	--	--	--	USR*	RSV	RM	--	--
N. Rough-winged Swallow	USR	CSR*	CSR	OSR*	USR*	USR*	CSR*	CSR*	CSR*	CSR*
Bank Swallow	CSR	OSR	USR	USR*	CSR*	USR*	USR*	CSR*	USR*	USR
Cliff Swallow	RSR*	OM	CSR*	CSR*	CSR	CSR*	ASR*	CSR*	USR*	USR
Barn Swallow	ASR*	ASR*	CSR*	ASR*	CSR*	CSR*	CSR*	CSR*	CSR*	CSR*
JAYS, MAGPIES, CROWS										
Steller's Jay	--	--	OWV	--	--	RWV	--	--	--	--
Blue Jay	USR*	OPR*	CPR*	CPR*	CPR*	UPR*	CSR*	CPR*	CPR*	CPR*
Gray Jay	--	--	--	--	--	RWV	--	--	--	--
Pinyon Jay	--	--	--	--	--	CPR*	--	--	--	--
Clark's Nutcracker	--	--	RSpV	--	--	RFV	RFV	RFV	--	--
Black-billed Magpie	OPR	CPR*	CPR*	UPR*	--	CPR*	UPR*	UPR*	OPR	--
American Crow	OPR	OPR*	CPR*	CPR*	CPR*	CPR*	CSR*	CPR*	CPR*	CPR*

Refuge, Region, or Area	C.L.R.	VAL.R.	F.N.R.	R.W.B.	D.S.R.	NW.Ne	N.P.V.	P.R.V.	LANC.	OMAHA
TITMICE										
Black-capped Chickadee	OPR	OPR*	CPR*	CPR*	CPR*	CPR*	UPR*	CPR*	CPR*	CPR*
Tufted Titmouse	--	--	--	--	OSR	--	--	RSpM	RPR*	CPR*
NUTHATCHES										
Red-breasted Nuthatch	UM	RM	--	OM	RM	UPR	UWM	UWM	UWM	UWM
White-breasted Nuthatch	OSR	RM	USR*	UPR*	CPR*	UPR	CPR*	CPR*	CPR*	CPR*
Pygmy Nuthatch	--	--	--	--	--	UPR*	--	--	--	--
CREEPERS										
Brown Creeper	OM	RM	USR	OM	OM	UPR	UWM	OWM	UWM	UPR
WRENS										
Rock Wren	RM	--	CSR*	--	--	USR*	CSR*	RFM	--	UM
Carolina Wren	--	--	--	--	--	--	--	RM	RPR	UPR
Bewick's Wren	--	--	--	--	--	--	--	RSpM	RM	--
House Wren	USR*	OSR*	CSR*	CSR*	CSR*	CSR*	CSR*	CSR*	CSR*	CSR*
Winter Wren	--	RM	--	UWM	RWM	RWM	RWM	RM	RWM	UWM
Sedge Wren	--	RSR	OSR	USR*	RSR*	--	--	CSR*	USR*	CSR
Marsh Wren	ASR*	ASR*	OSR	USR	OM	USR*	CSR*	USR*	USR*	CSR
DIPPERS										
American Dipper	--	--	--	--	--	RM	--	--	--	--
KINGLETS, THRUSHES										
Golden-crowned Kinglet	OM	--	--	CM	OM	RWM	UWM	UWM	UWM	UWM
Ruby-crowned Kinglet	UM	RM	UM	UM	UM	UFM	UM	UM	UWM	CM
Blue-gray Gnatcatcher	--	--	--	--	--	--	UM	--	RSR	USR
Eastern Bluebird	OSR	RSR*	USR*	UPR*	OPR*	RSR	USR*	CPR*	UPR*	CSR*
Mountain Bluebird	UM	OSR*	USR	--	--	CSR*	UWM	OM	RM	--
Townsend's Solitaire	UM	RM	OM	OWM	OWM	CWM	UWM	OM	RM	--
Veery	OM	OM	--	OM	--	RSM	RSpM	OSpM	RM	OM
Gray-cheeked Thrush	USpM	OM	OM	USpM	RSpM	RM	--	OSpM	UM	UM
Swainson's Thrush	UM	OM	OM	CSpM	RSpM	UM	CSpM	UM	CM	UM
Hermit Thrush	RM	--	--	--	RM	RSM	--	RFM	RM	UM
Wood Thrush	OSpM	RM	OM	USR*	USR*	RM	RSpM	--	OSR	USR*
American Robin	CPR*	CPR*	CPR*	CPR*	CPR*	CPR*	CPR*	CPR*	CPR*	CSR*
Varied Thrush	--	--	--	--	--	--	--	--	--	RWV
MIMIC THRASHERS										
Gray Catbird	RSR	OSR*	USR*	CSR*	CSR*	RSR	USR*	USR*	CSR*	CSR*
Northern Mockingbird	OSR	RSR	USR*	USR*	RSpM	RSR*	USR*	RSR*	RSR*	OM
Sage Thrasher	--	--	--	--	--	RSR*	--	--	--	--
Brown Thrasher	USR*	CSR*	CSR*	CSR*	CSR*	CSR*	CSR*	CSR*	CSR*	USR
PIPITS										
Water Pipit	RM	CM	UM	CM	OM	CM	UM	OM	UM	--
Sprague's Pipit	--	--	--	RM	--	--	--	OM	RM	--
WAXWINGS										
Bohemian Waxwing	OM	--	UM	OWM	RWM	CWM	UWM	RWM	RWM	RWM
Cedar Waxwing	UPR	OM	UM	UPR	OPR*	OPR*	CPR*	UPR*	UPR	UM

Refuge, Region, or Area	Federal Refuges or Managed Areas					Non-Federal Lands or Regions				
	C.L.R.	VAL.R.	F.N.R.	R.W.B.	D.S.R.	NW.Ne	N.P.V.	P.R.V.	LANC.	OMAHA
SHRIKES										
Northern Shrike UM	OM	OM	UM	RWM	UWM	UWM	OWM	RWM	RWM	
Loggerhead Shrike	CSR*	OSR	CSR	CPR*	OPR*	CSR*	CSR*	UPR*	UPR*	OSR
STARLINGS										
European Starling	UPR*	OPR*	CPR*	APR*	CPR*	CPR*	CPR*	APR*	ASPR*	CPR*
VIREOS										
White-eyed Vireo	--	--	--	--	--	--	--	--	RSpM	OM
Bell's Vireo	RSR*	OSR*	USR	CSR*	USR*	USR	CSR*	CSR*	CSR*	RM
Plumbeous Vireo	OSR*	--	--	--	RM	RSR*	RM	OSpM	RM	UM
Yellow-throated Vireo	--	--	--	--	--	--	--	RSpM	RM	CSR
Warbling Vireo	RSR*	CSR*	--	CSR*	CSR*	CSR*	CSR*	CSR*	USR*	CSR
Philadelphia Vireo	--	--	--	UM	RM	--	--	OSpM	UM	UM
Red-eyed Vireo	OSR	CSR*	USR*	OSR*	OSR*	USR*	USR*	RSR	USR	CSR*
WOOD WARBLERS										
Blue-winged Warbler	--	RM	--	--	--	--	--	RSpM	RSpM	UM
Golden-winged Warbler	--	--	--	RSpM	--	--	RSpM	RSpM	RM	UM
Tennessee Warbler	OSpM	--	OM	CM	UM	RM	RSpM	USpM	CM	CM
Orange-crowned Warbler	UM	OM	CM	CM	UM	UM	UM	CM	CM	CM
Nashville Warbler	--	OM	--	UM	OM	RM	RM	OM	UM	CM
Northern Parula	OSpM	--	--	RM	OSpM	RM	--	--	RM	USR
Yellow Warbler	USR*	CSR*	CSR*	CSR*	CSR*	CSR*	CSR*	CSR*	CSR*	USR
Chestnut-sided Warbler	--	--	--	UM	OSpM	RSM	RSpM	USpM	UM	UM
Magnolia Warbler	OM	--	--	UM	OM	RM	--	USpM	RM	UM
Cape May Warbler	--	--	--	--	--	RSM	--	RSpM	RSpM	UM
Black-thrd. Blue Warbler	OFM	--	--	RSpM	--	--	--	RSpM	RM	UM
Yellow-rumped Warbler	CM	CM	CM	CM	CM	USR*	CM	CM	CM	CM
Townsend's Warbler	OM	--	--	--	--	RFM	--	--	--	--
Black-thrd. Green Warbler	--	--	--	USpM	OSpM	RFM	--	RM	--	UM
Blackburnian Warbler	--	RM	--	UM	RSpM	--	--	OSpM	RM	UM
Yellow-throated Warbler	--	--	--	RSpM	--	RSM	--	RSpM	--	UM
Pine Warbler	--	--	--	--	--	--	--	--	RM	OM
Palm Warbler	--	RM	--	UM	OSpM	RM	--	OSpM	RM	UM
Bay-breasted Warbler	--	--	--	UM	OSpM	--	--	RSpM	RM	UM
Blackpoll Warbler	UM	RM	OM	CM	USpM	RM	USpM	USpM	CM	UM
Cerulean Warbler	--	--	--	--	--	RSM	--	--	--	UM
Black-and-white Warbler	OM	RM	USR	CSpM	OSR*	USR	USpM	USpM	UM	UM
American Redstart	UM	OSR*	USR*	CSR	CSR*	CSR*	UM	UM	UM	CSR*
Prothonotary Warbler	--	--	--	--	--	--	RSpM	--	--	USR
Worm-eating Warbler	OSpM	--	--	--	--	--	--	--	--	OM
Ovenbird	UM	RSR*	OSR*	UM	USR*	CSR	OSR	OSR	RM	CSR*
Northern Waterthrush	OM	--	--	UM	RSpM	RM	RSpM	OSpM	UM	UM
Louisiana Waterthrush	--	--	--	--	--	--	--	RSpM	RSpM	UM
Kentucky Warbler	--	--	--	--	--	--	--	RSpM	RM	UM
Connecticut Warbler	--	--	RM	RSpM	--	--	--	RSpM	RM	UM
Mourning Warbler	--	--	--	USpM	--	--	RSpM	RM	RM	UM
MacGillivray's Warbler	OM	RM	--	--	--	RM	UM	--	--	--
Common Yellowthroat	CSR*	CSR*	CSR*	CSR*	CSR*	CSR	CSR*	CSR*	CSR*	CSR*
Hooded Warbler	--	--	--	RSpM	RSpM	RSM	RSpM	--	RSpM	UM
Wilson's Warbler	UM	RM	UM	UM	OM	UM	UM	OM	UM	UM
Canada Warbler	--	OM	--	UM	OM	--	--	--	UM	UM
Yellow-breasted Chat	OSR	OSR	CSR*	OSR	--	USR*	USR*	RSpM	RM	RSR

Refuge, Region, or Area	Federal Refuges or Managed Areas					Non-Federal Lands or Regions				
	C.L.R.	VAL.R.	F.N.R.	R.W.B.	D.S.R.	NW.Ne	N.P.V.	P.R.V.	LANC.	OMAHA
TANAGERS										
Summer Tanager	--	--	--	--	--	--	--	RSpM	RSR	OSR
Scarlet Tanager	--	--	USR*	USpM	--	RSV	RM	RSR	RSR	USR
Western Tanager	OSR	OSR	USR	--	--	USR*	RM	--	--	RM
GROSBEAKS, BUNTINGS										
Northern Cardinal	--	RM	UPR	CPR	CPR*	OV	UPR*	CPR*	CPR*	CPR
Rose-breasted Grosbeak	OSR	--	OSR	USR*	CSR*	RM	USR	USR*	CSR*	CSR
Black-headed Grosbeak	USR	--	CSR*	--	--	CSR*	CSR*	OSR*	RSR*	--
Blue Grosbeak	USR*	OSR*	USR*	USR*	RSR	RSR*	USR	USR*	RSR*	--
Lazuli Bunting	RSR	OSR*	UM	USpM	--	USR*	USR	RSpM	RSpM	RM
Indigo Bunting	--	OM	UM	USR*	USR*	USR	USR*	USR*	CSR*	CSR*
Dickcissel	RSR*	CSR*	CSR*	CSR*	CSR*	USR	USR*	CSR*	CSR*	USR*
NEW WORLD SPARROWS										
Green-tailed Towhee	--	--	--	--	--	RSM	--	--	--	--
Eastern & Spotted Towhee	UM	CSR*	ASR*	CM	CSR*	CSR*	CSR*	CSR*	RSR	CSR*
American Tree Sparrow	UM	AWM	UWM	CWM	AWM	AWM	CWM	CWM	CWM	CWM
Chipping Sparrow	RSR	CSR*	CSR*	CM	USR	CSR*	USR*	CSR*	RSR*	USR*
Clay-colored Sparrow	RM	UM	UM	CM	USpM	CM	CM	CM	OM	UM
Brewer's Sparrow	--	--	--	--	--	CSR*	--	--	--	--
Field Sparrow	RM	OSR*	USR*	USR	CSR*	RSR	CSR*	CSR*	USR*	USR*
Vesper Sparrow	RSR*	CSR*	ASR*	CM	USR*	USR*	CM	CM	UM	UM
Lark Sparrow	CSR*	CSR*	ASR*	CSR*	USR*	CSR*	CSR*	CSR*	UM	UM
Lark Bunting	CSR*	CSR*	CSR	CSR*	--	CSR*	CSR	OSR*	RSpM	--
Savannah Sparrow	RSR	OSR	USR*	--	USR	RSR	UM	CM	UM	UM
Baird's Sparrow	--	OM	RM	UM	RSpM	RM	--	--	RM	--
Grasshopper Sparrow	ASR*	CSR*	ASR*	CSR*	CSR*	CSR*	CSR*	CSR*	USR*	USR
Henslow's Sparrow	--	--	--	OM	RSR*	--	--	RM	RSR*	--
Le Conte's Sparrow	--	OM	--	--	UM	--	--	RM	RM	UM
Sharp-tailed Sparrow	--	RM	--	--	--	--	--	RM	RM	--
Fox Sparrow	--	--	RM	OM	UM	--	--	RM	RM	UM
Song Sparrow	RPR	OPR*	UPR	CM	CPR*	UPR	UPR*	CPR*	CPR*	CPR
Lincoln's Sparrow	RM	RM	UM	CM	UM	UM	UM	UM	UM	CM
Swamp Sparrow	OSR	--	--	RSR*	UM	CSR	USR*	OM	UM	UM
White-throated Sparrow	RM	UM	UM	UM	UM	UM	UM	UM	CWM	CM
White-crowned Sparrow	CM	UM	UM	CM	UM	CM	CM	UWM	UWM	CM
Harris' Sparrow	UM	OM	UM	CM	CM	CWM	UM	CWM	CWM	CWM
Dark-eyed Junco	CWM	CWM	CWM	CWM	AWM	CPR*	CWM	CWM	CWM	CWM
McCown's Longspur	RM	OM	UM	--	--	CSR*	--	--	--	--
Chestnut-collared Longspur	RM	OWM	UWM	AWM	--	CSR	RSpM	--	--	--
Lapland Longspur	RWM	OWM	UWM	AWM	--	AWM	UWM	OWM	RWM	RM
Snow Bunting	OWM	RWM	--	RWM	RWM	AWM	RWM	--	RWM	--
BLACKBIRDS, ORIOLES										
Bobolink	CSR*	CSR*	UM	CSR	OSR*	CSR*	USR*	CSR*	RSR*	--
Red-winged Blackbird	APR*	ASR*	CSR*	CPR*	APR*	APR*	ASR*	APR*	APR*	CSR*
Eastern Meadowlark	CSR*	CPR*	USR*	OSR*	CPR*	CSR*	USR	USR*	CPR*	OPR*
Western Meadowlark	APR*	CPR*	APR*	CSR*	CPR*	CPR*	CSR*	CPR*	CPR*	CPR*
Yellow-headed Blackbird	ASR*	CSR*	CSR	CM	UM	CSR*	USR*	CSR*	USR	USR
Rusty Blackbird	--	--	OM	CWM	OWM	RM	OM	OM	UWM	RSR
Brewer's Blackbird	RM	OM	UM	CM	OM	CSR*	UWM	OM	UWM	--

Refuge, Region, or Area	Federal Refuges or Managed Areas					Non-Federal Lands or Regions				
	C.L.R.	VAL.R.	F.N.R.	R.W.B.	D.S.R.	NW.Ne	N.P.V.	P.R.V.	LANC.	OMAHA
Great-tailed Grackle	--	--	--	USR	--	--	USR*	CPR*	CSR*	RSR
Common Grackle	CPR*	CSR*	CSR*	APR*	CPR*	CPR*	CPR*	CPR*	APR*	CSR*
Brown-headed Cowbird	CSR*	CSR*	CSR*	CSR*	CPR*	CSR*	CSR*	CSR*	CSR*	CSR*
Orchard Oriole	CSR*	USR*	CSR*	CSR*	CSR*	CSR*	CSR*	CSR*	CSR*	USR*
Bullock's/Baltimore Oriole	USR*	CSR*	USR*	CSR*	CSR*	CSR*	CSR*	CSR*	CSR*	CSR*
FINCHES										
Rosy Finch	--	--	--	--	--	RWM	--	RWM	--	--
Pine Grosbeak	--	--	--	--	--	RWM	--	RWM	RWM	--
Purple Finch	OM	--	--	UWM	OM	CWM	UWM	OWM	RWM	UWM
Cassin's Finch	--	--	--	--	--	UWM	--	--	--	--
House Finch	--	--	--	--	--	RPR	UPR*	CPR*	CPR*	CPR*
Red Crossbill	OFM	RM	RWM	OWM	--	UPR*	--	OWM	RWM	RWM
Common Redpoll	OWM	OWM	UWM	OWM	RWM	CWM	OWM	RWM	RWM	RWM
Pine Siskin	OPR	OM	CSR*	OPR*	OWM	CPR*	UPR	CPR*	UPR*	UWM
American Goldfinch	CPR*	USR*	CSR*	CPR*	CPR*	CPR*	CPR*	CPR*	CPR*	CPR
Evening Grosbeak	OWM	RM	UWM	RWM	--	CWM	RWM	RWM	RWM	RWM
OLD WORLD SPARROWS										
House Sparrow	OPR*	OPR*	APR*	APR*	APR*	APR*	APR*	CPR*	APR*	CPR*

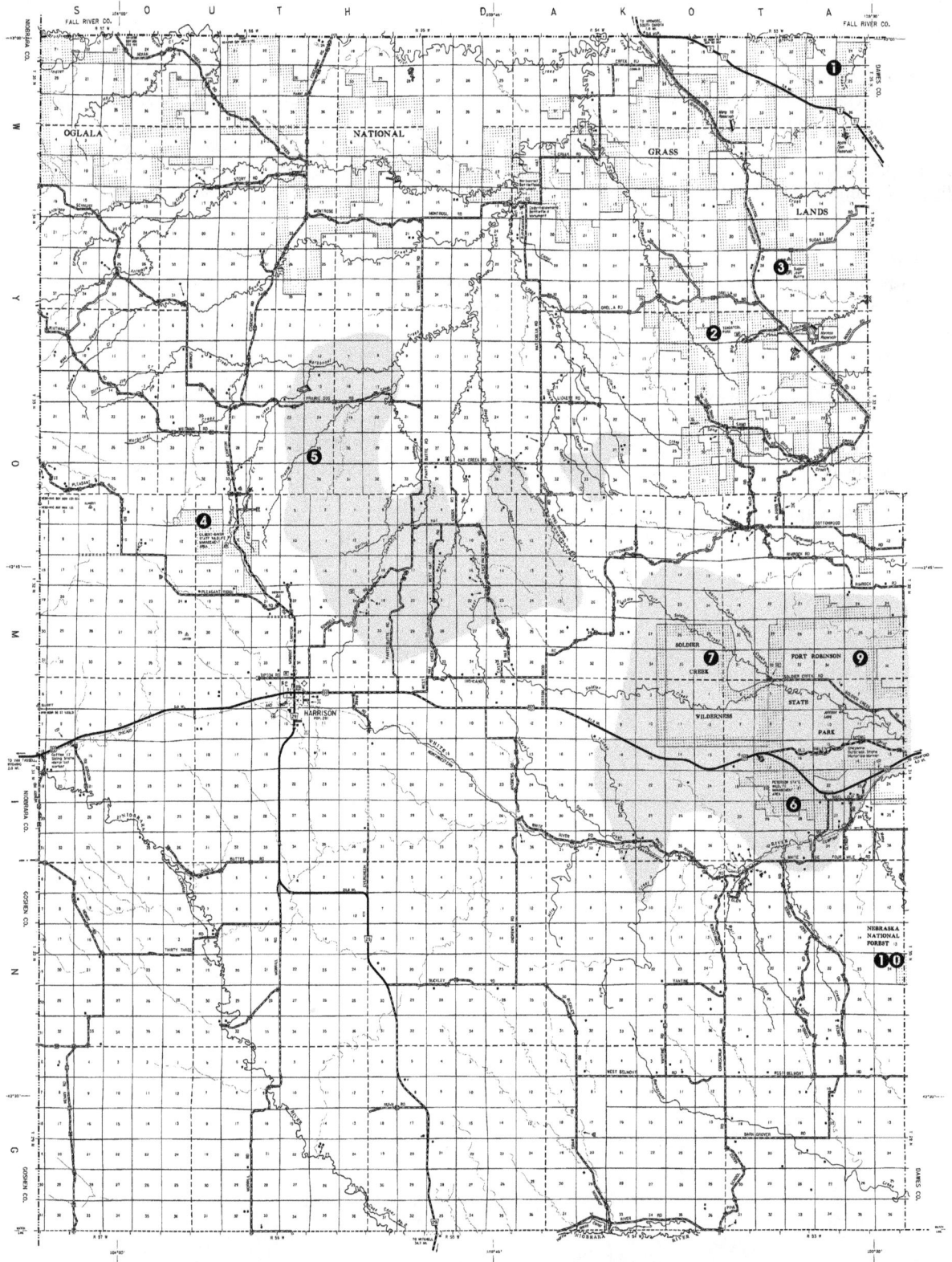

Sioux County (northern half)   Map 1

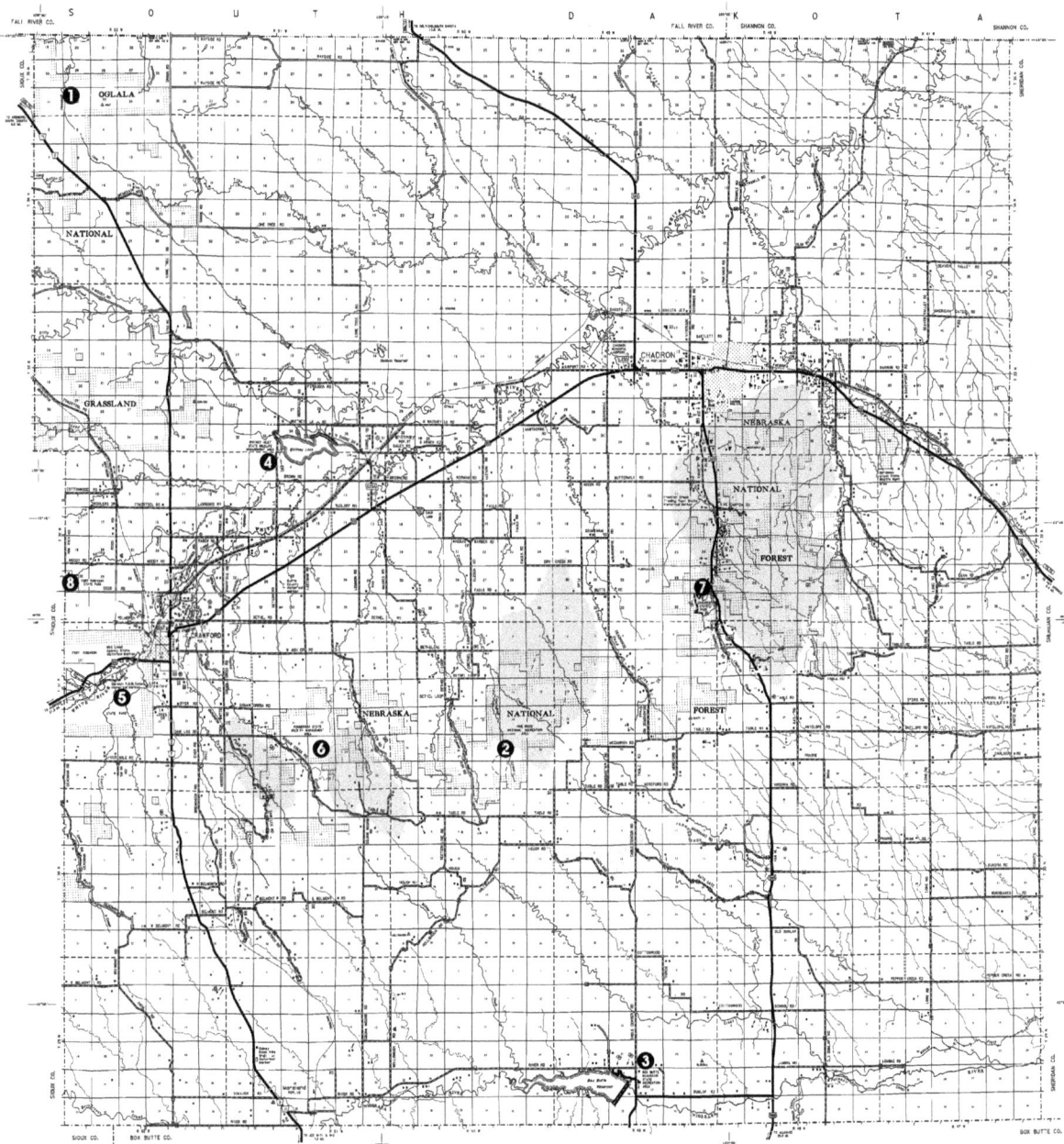

Dawes County (north and south maps)   Map 2

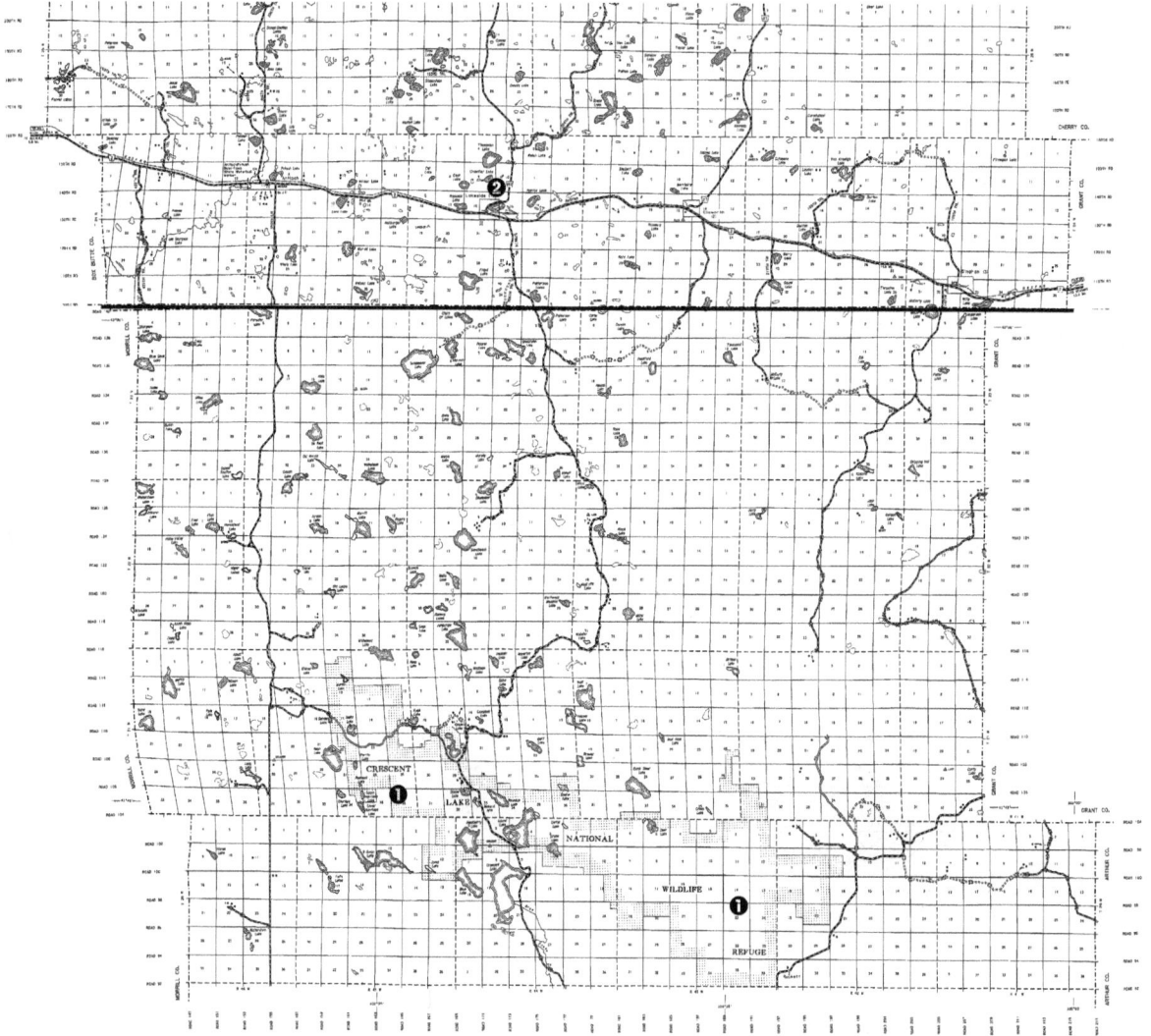

Garden County and Sheridan County

Map 3

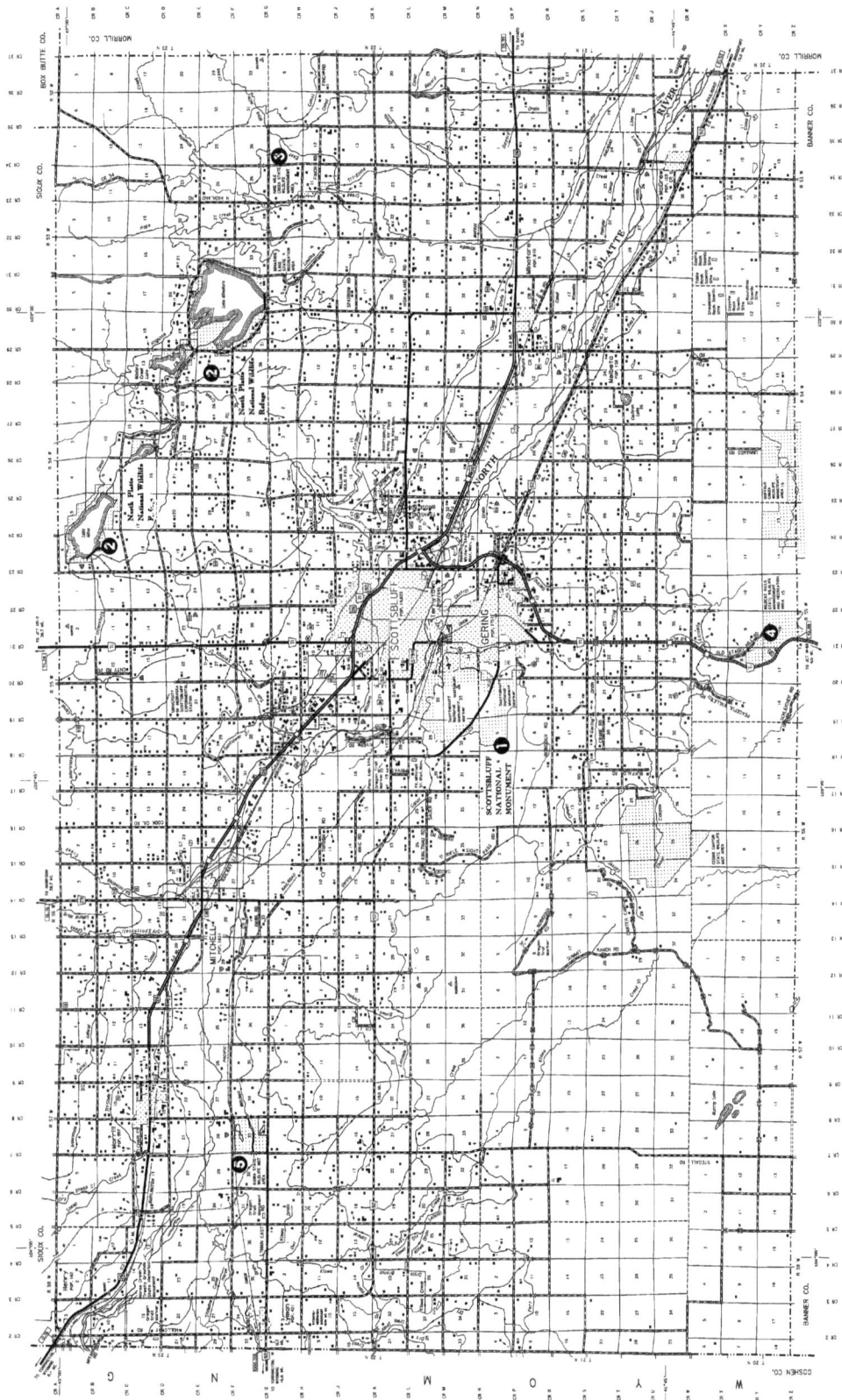

Scotts Bluff County

Map 4

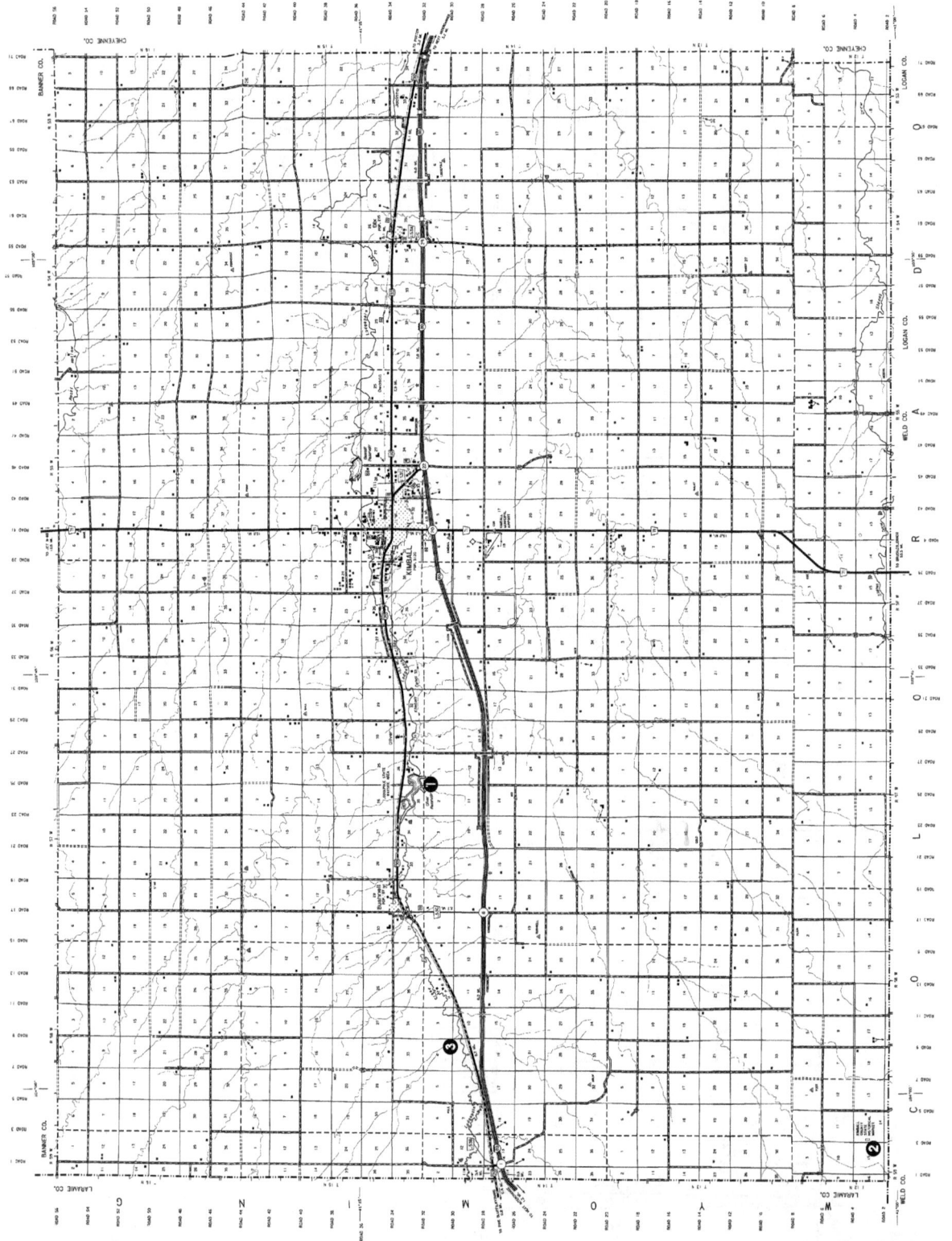

Kimball County

Map 5

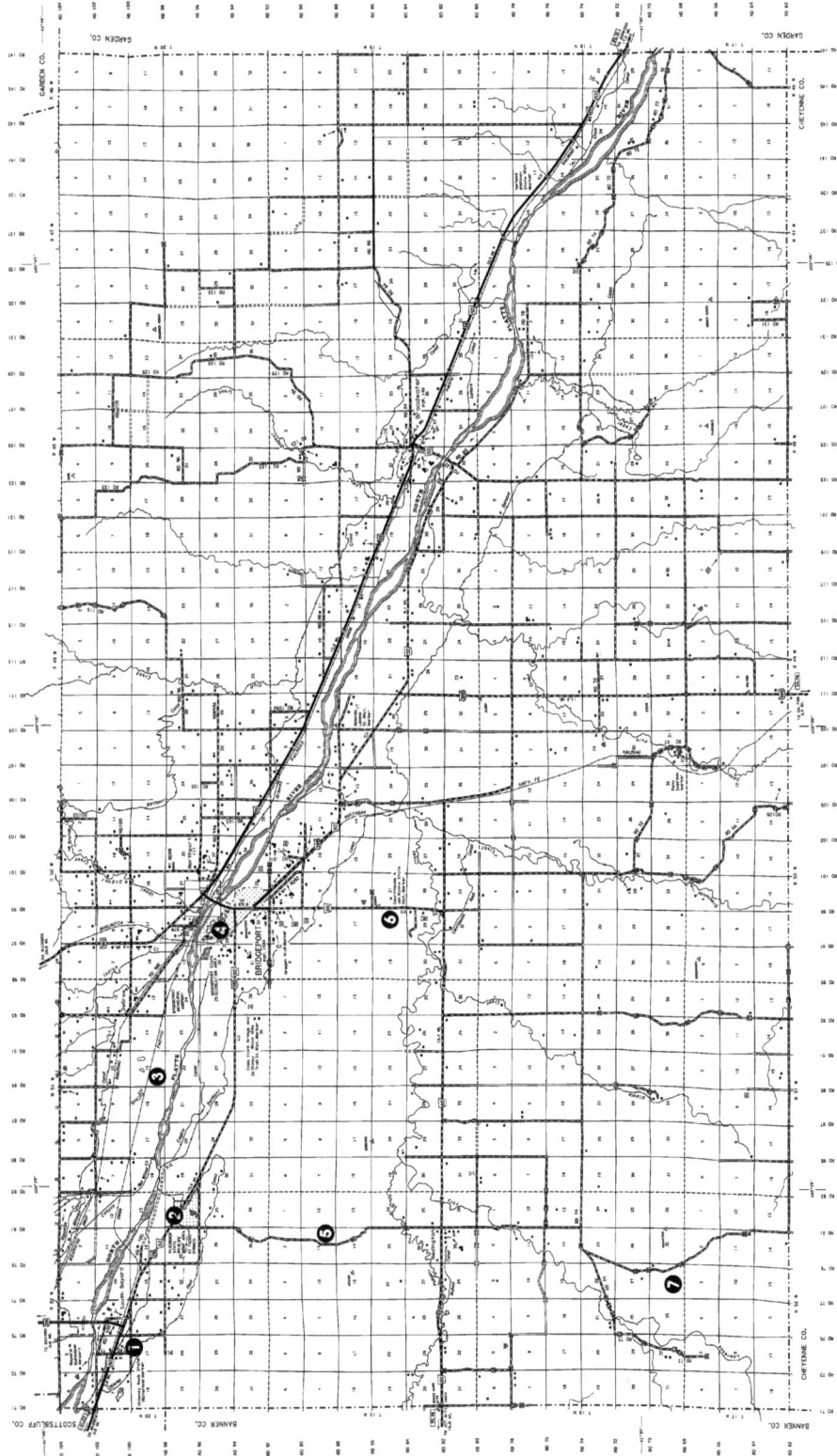

Morrill County

Map 6

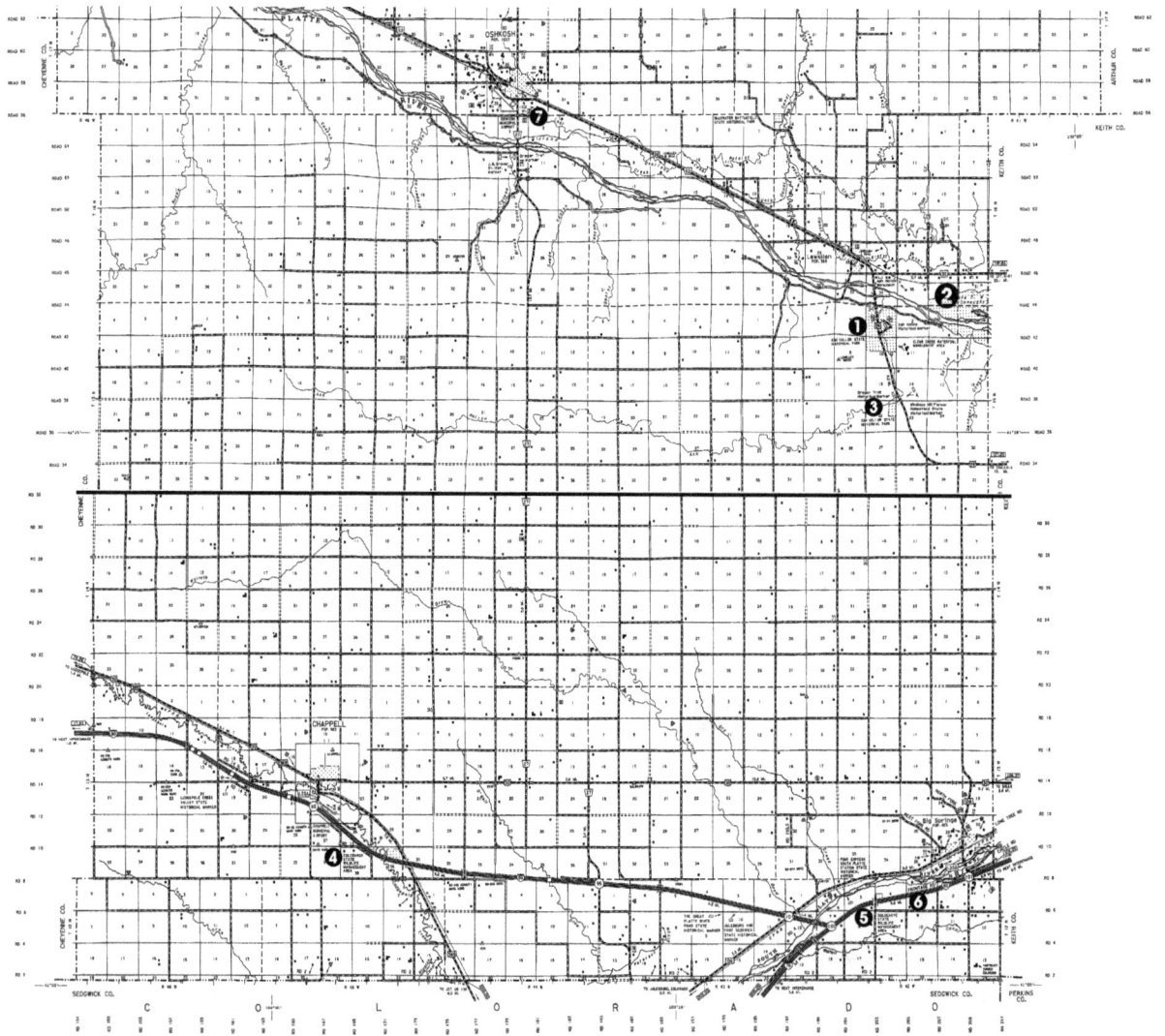

Garden and Deuel Counties

Map 7

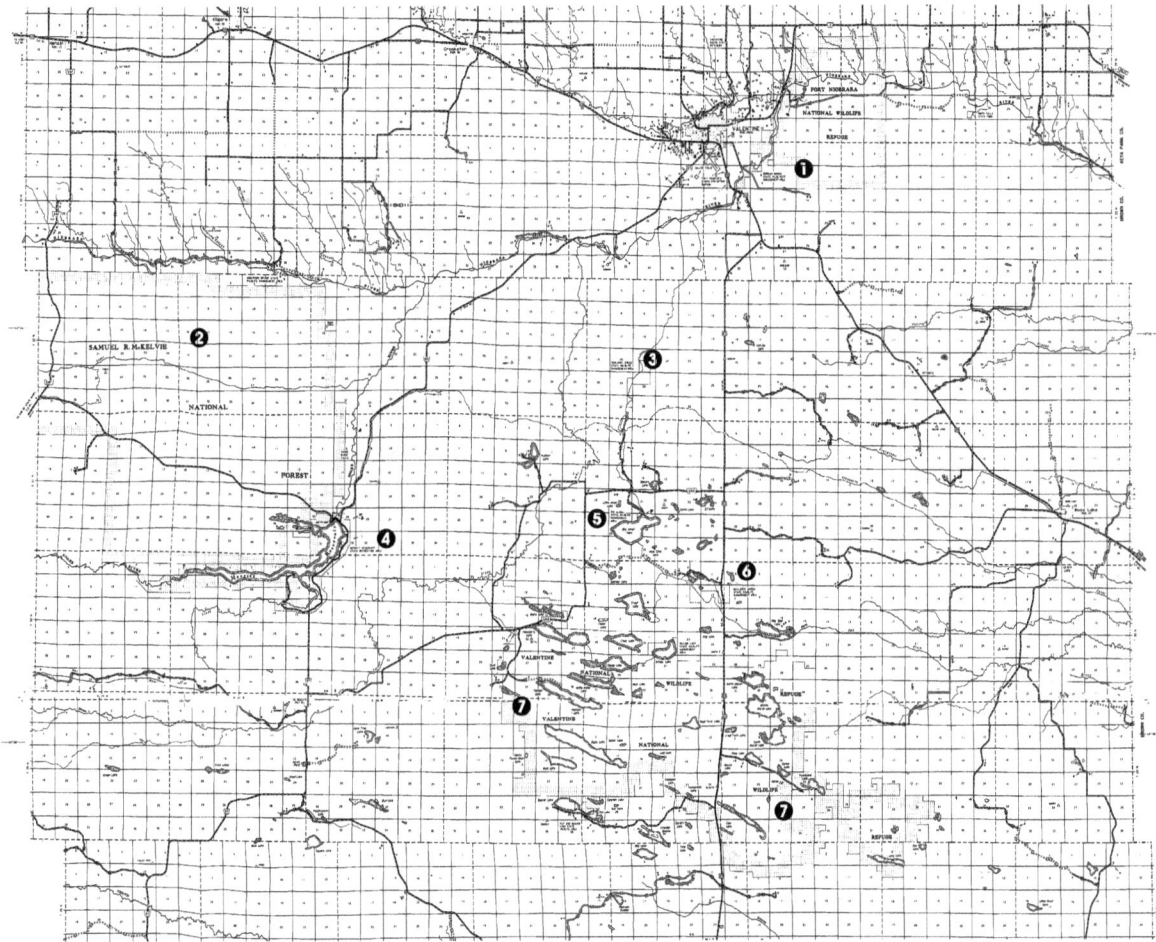

Cherry County   Map 8

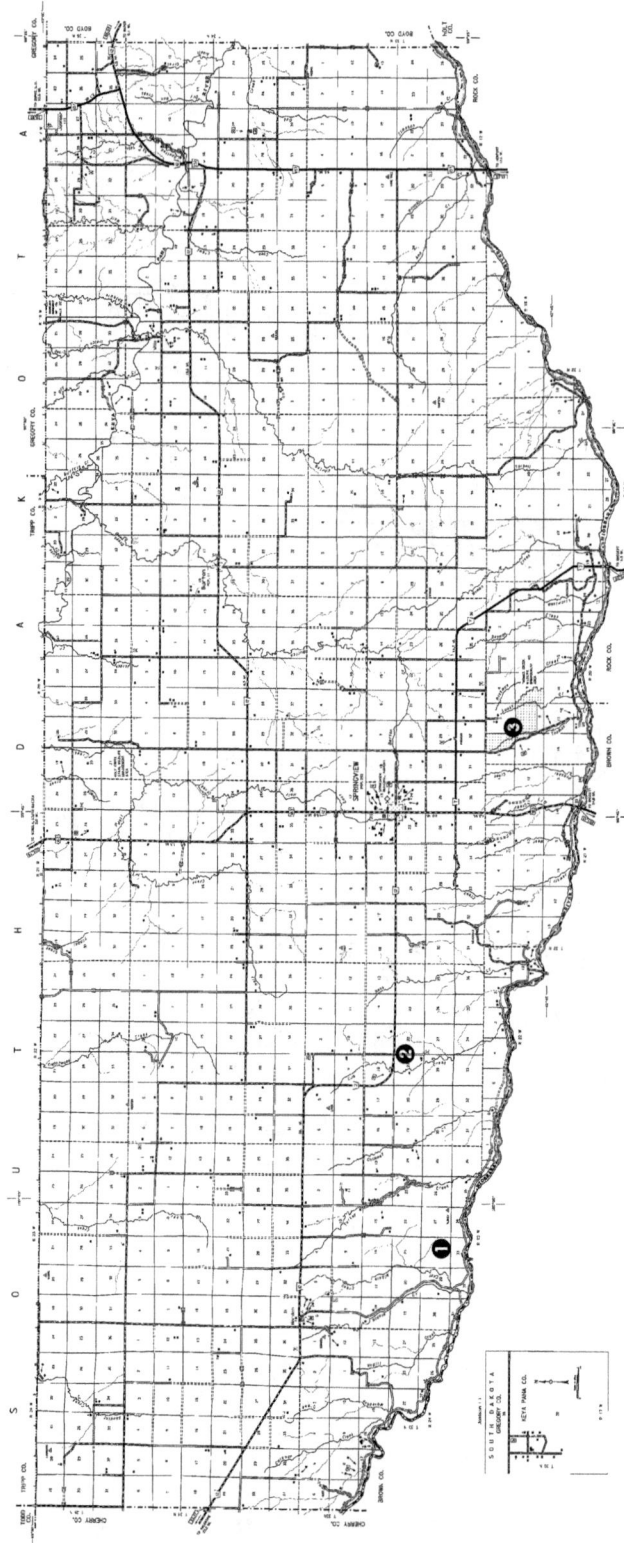

Keya Paha County

Map 9

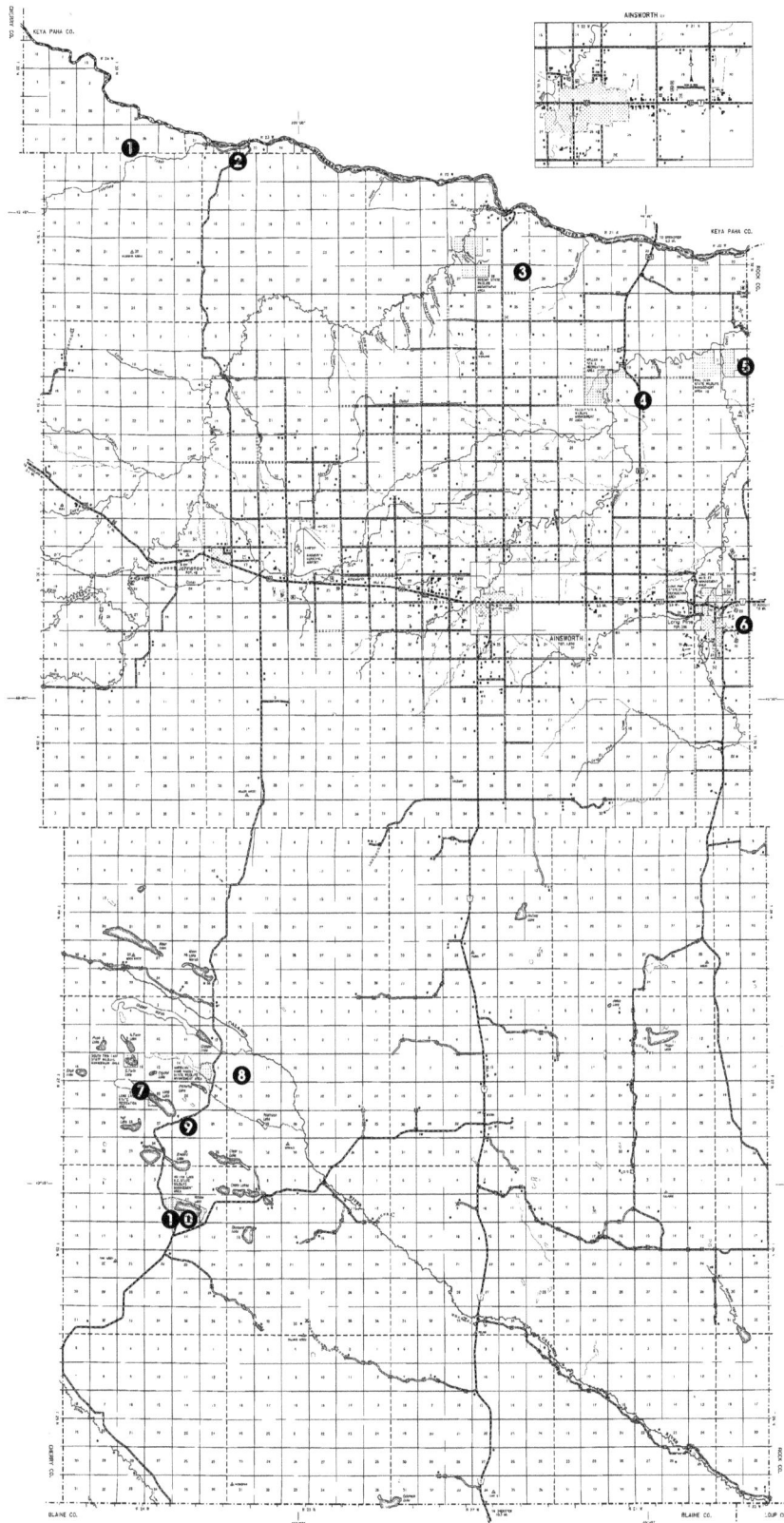

Brown County

Map 10

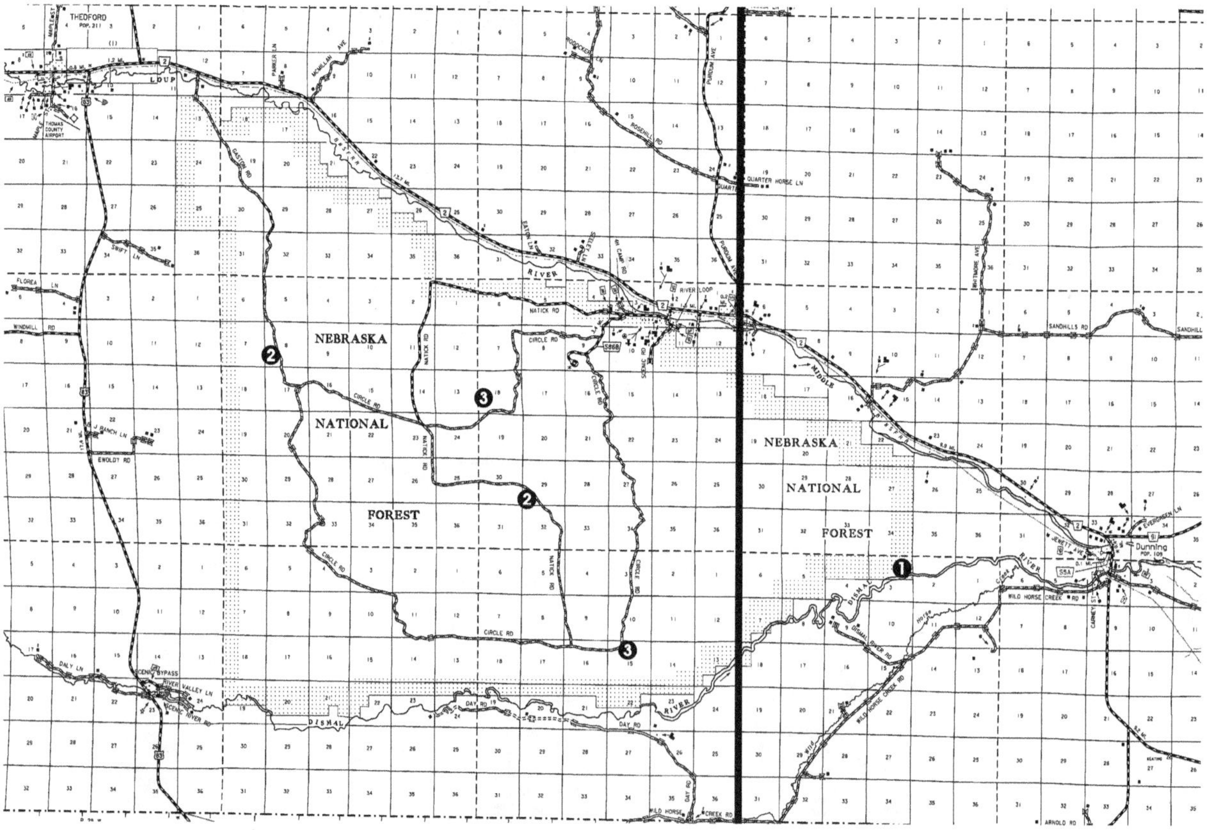

Thomas & Blaine Counties

Map 11

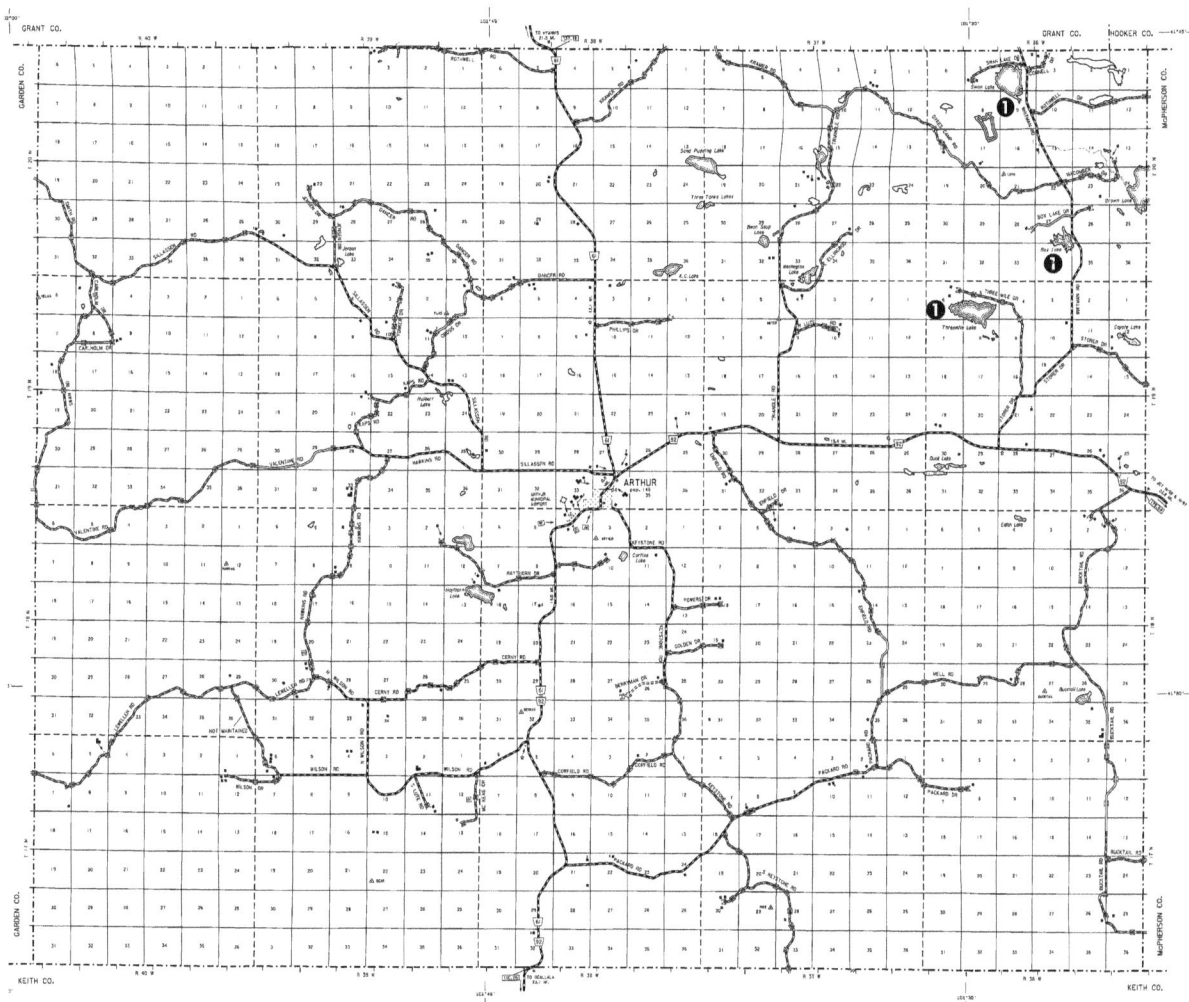

Arthur County   Map 12

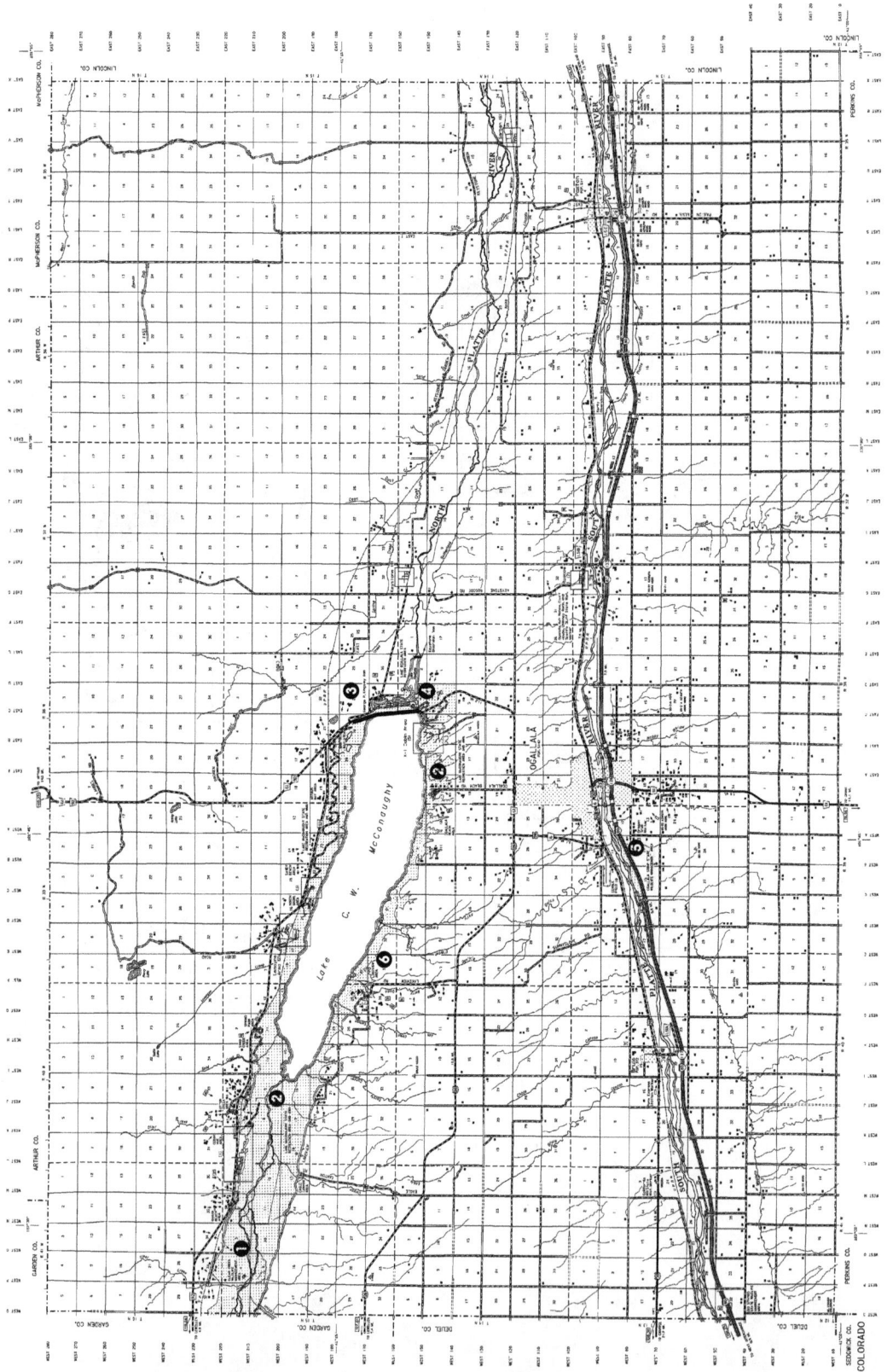

Keith County

Map 13

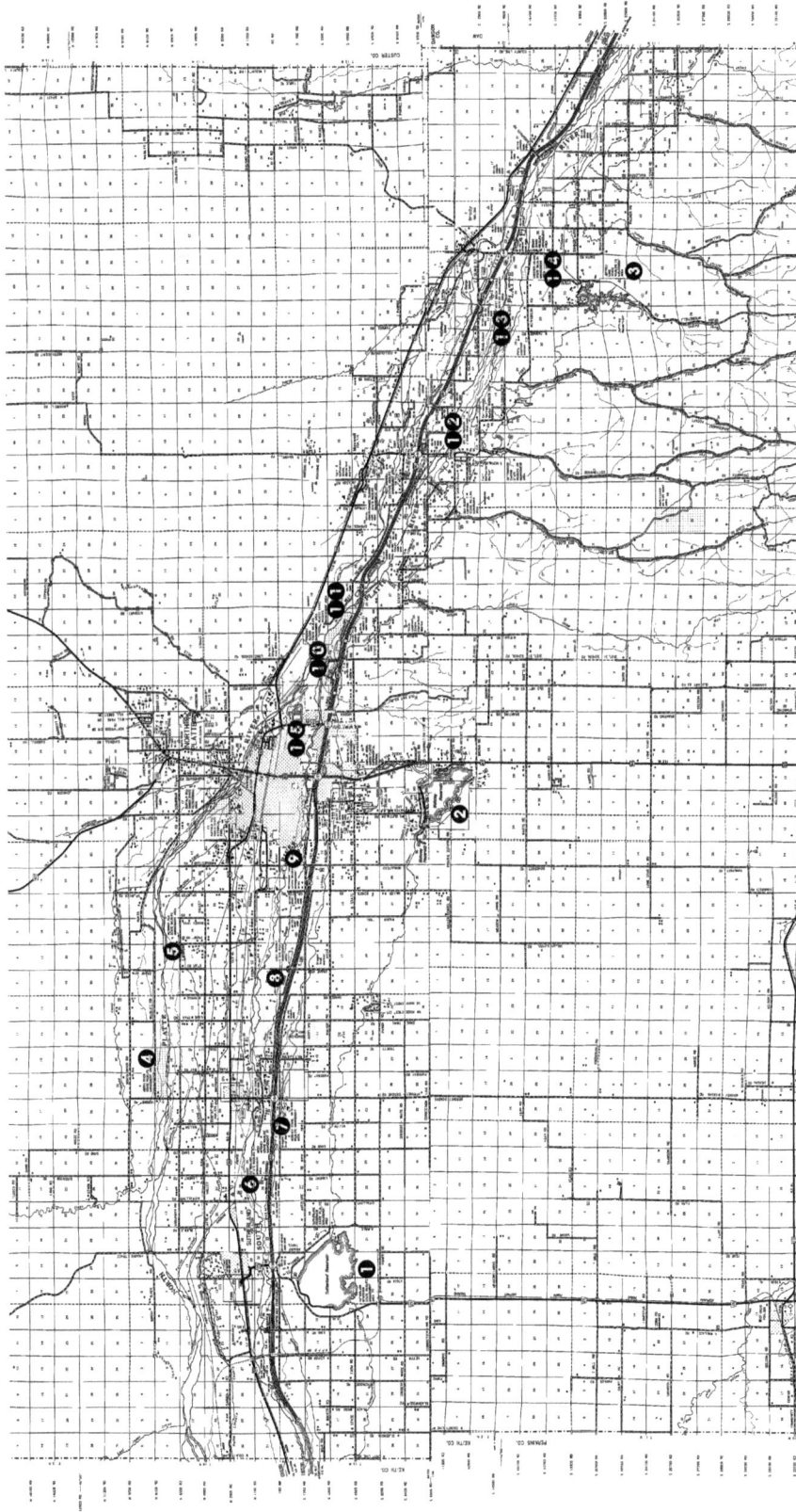

Lincoln County   Map 14

138

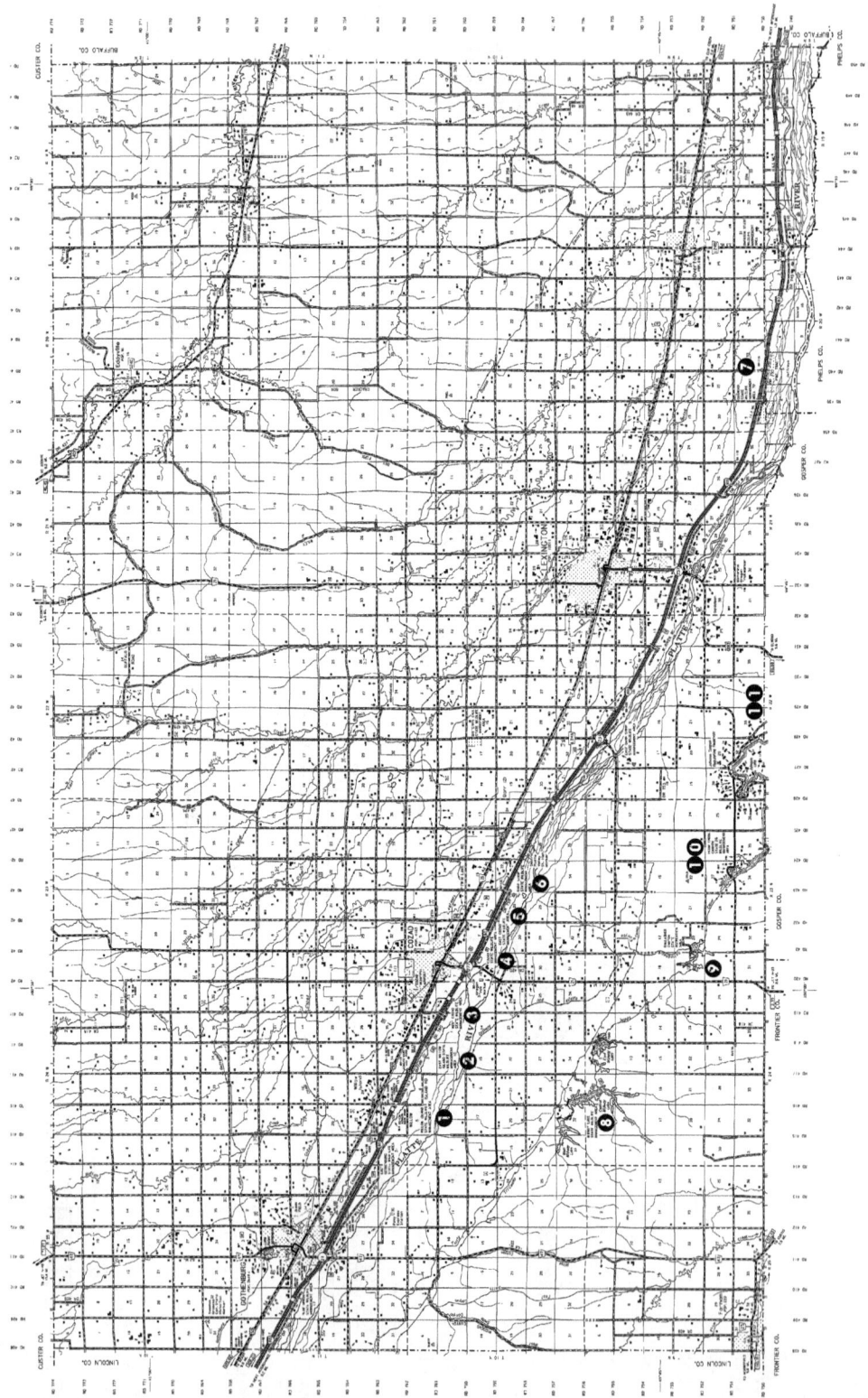

Dawson County

Map 15

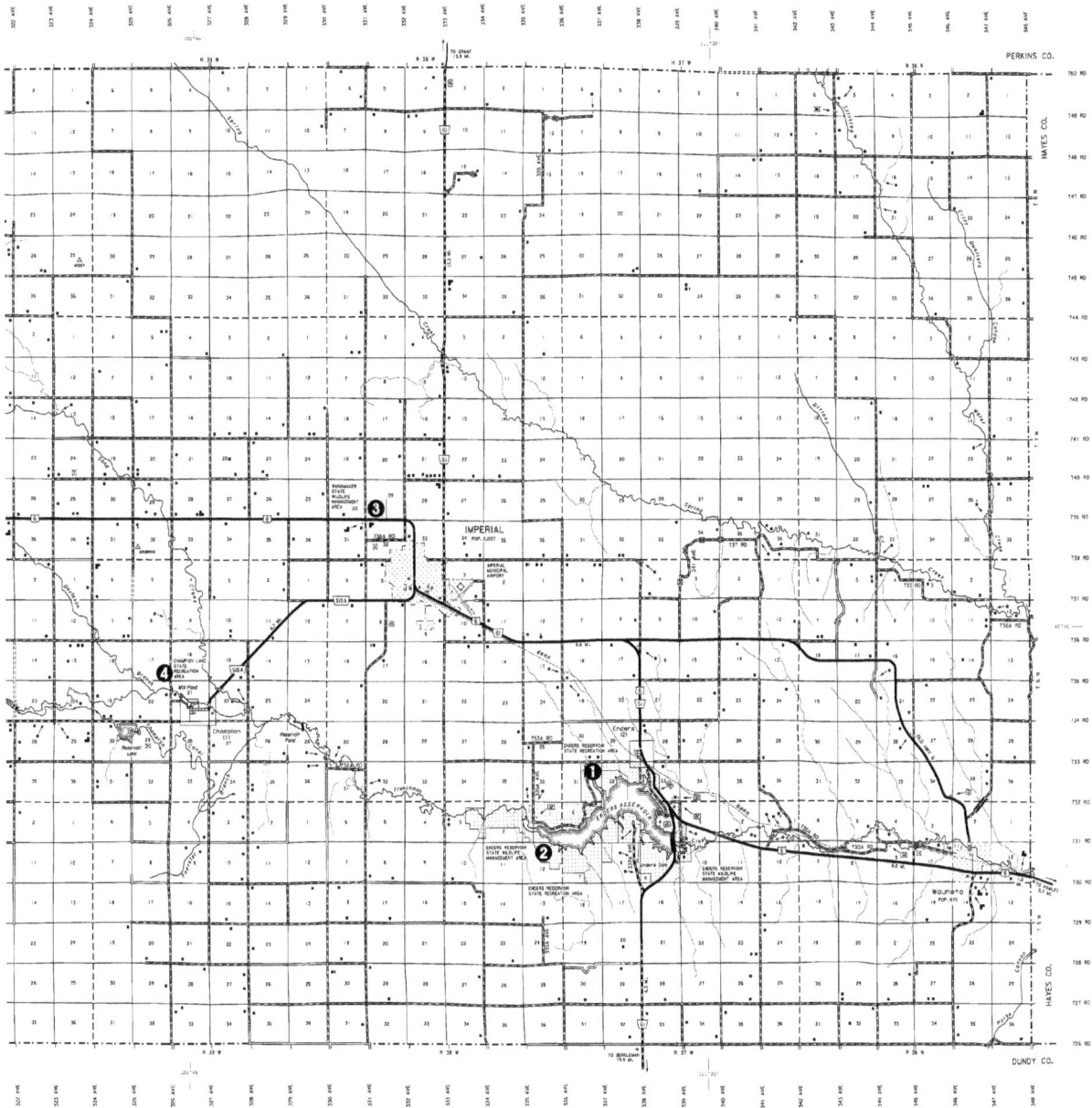

Chase County (selection)   Map 16

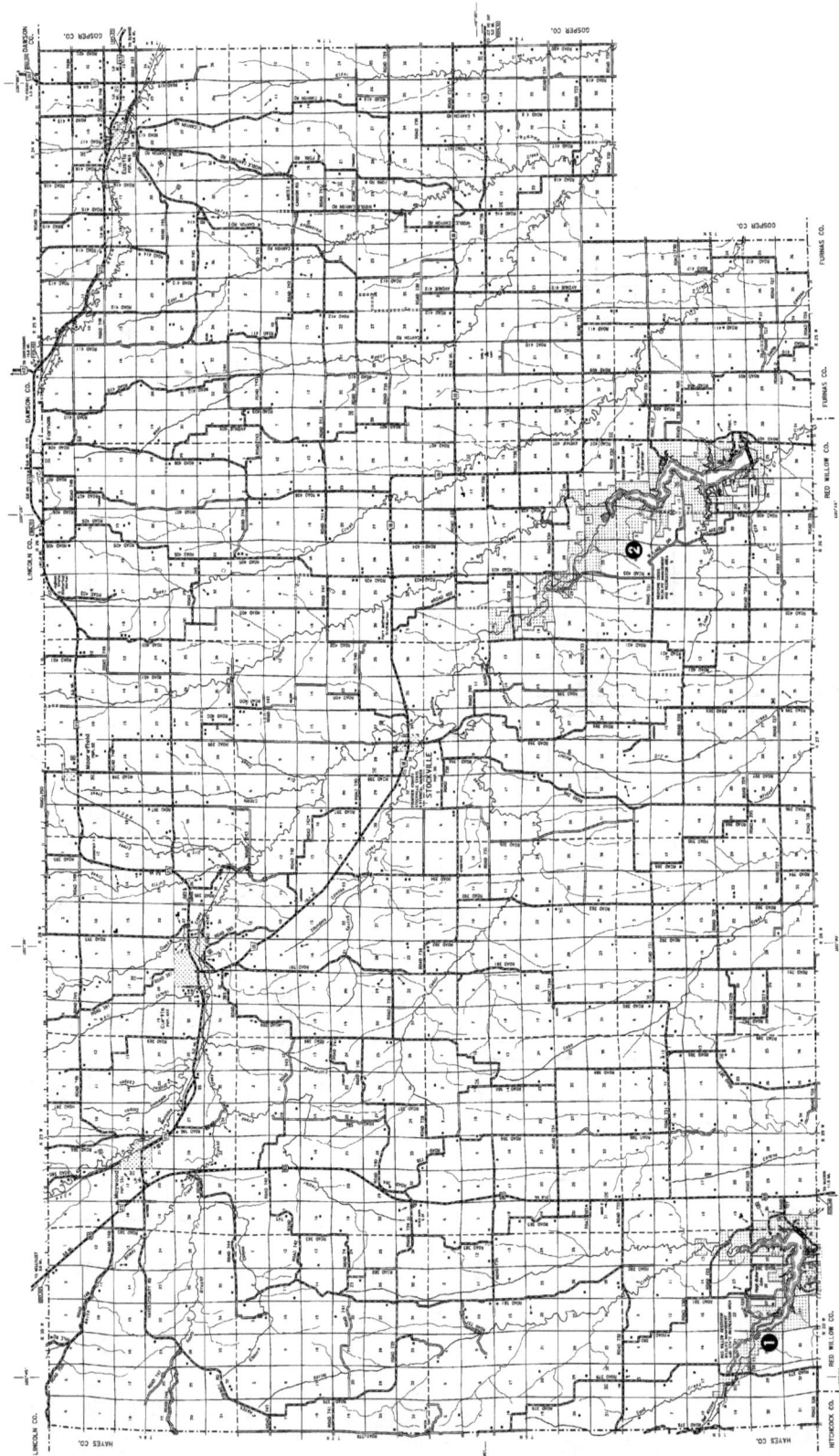

Frontier County

Map 17

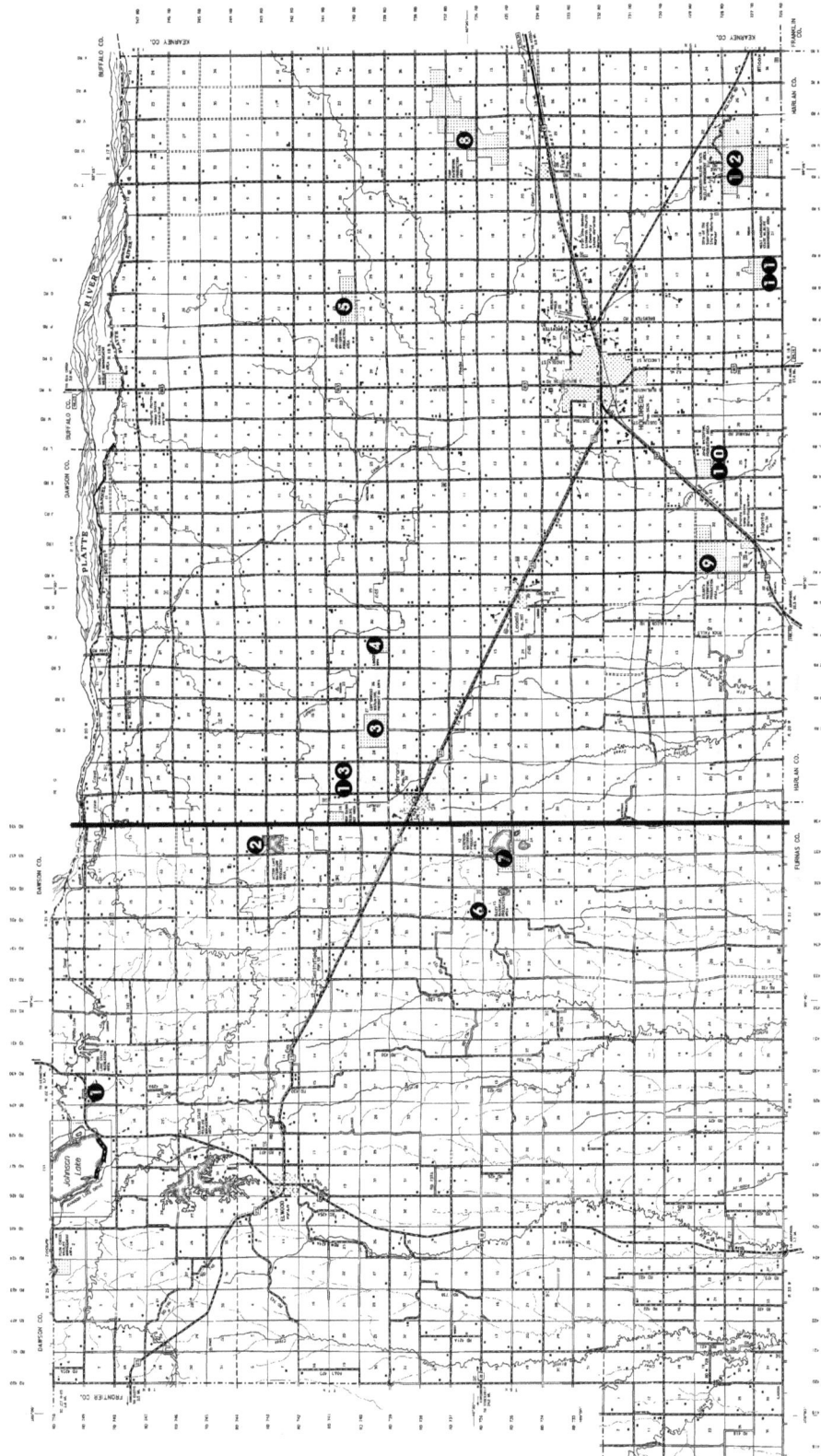

Gosper and Phelps Counties

Map 18

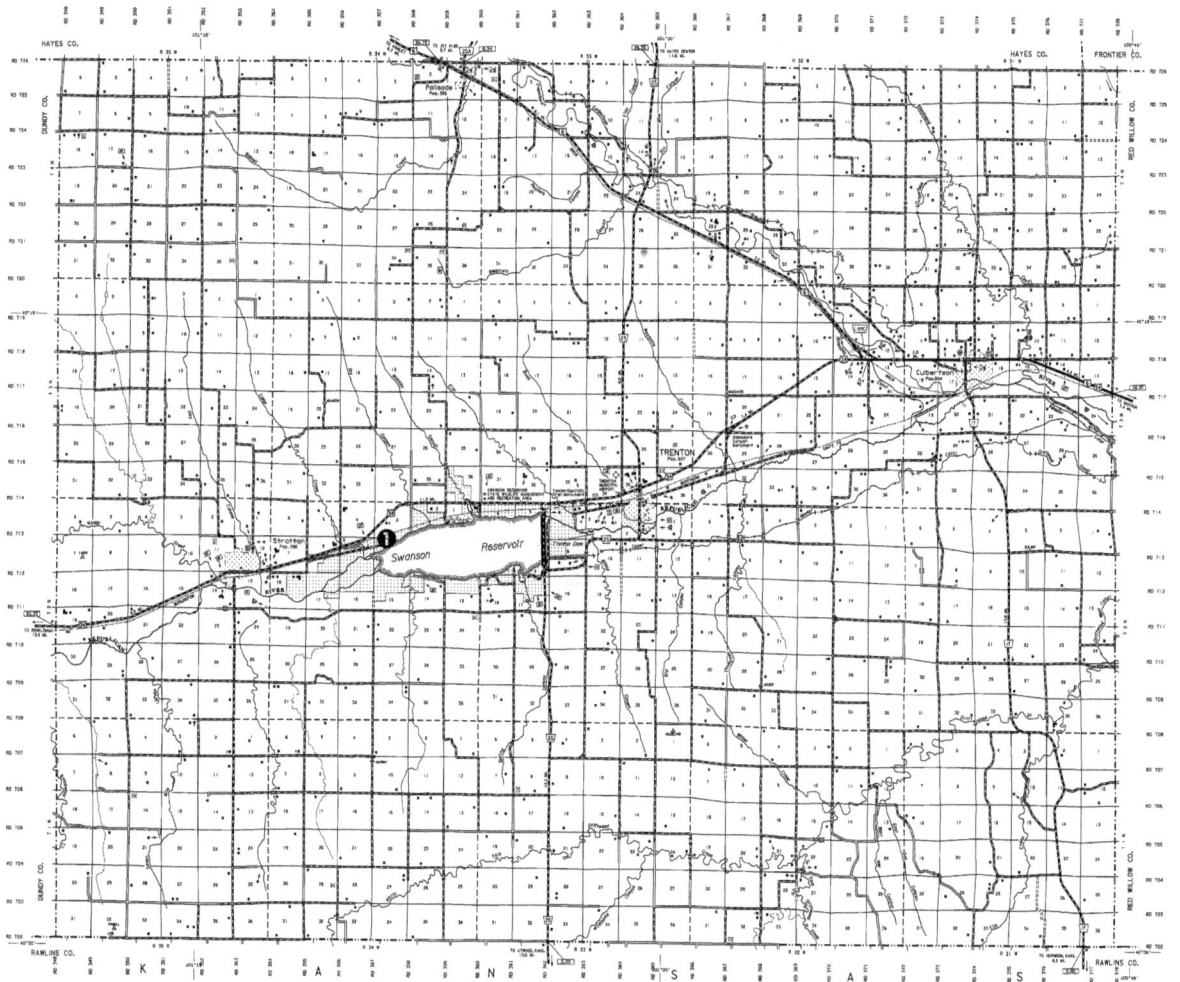

Hitchcock County

Map 19

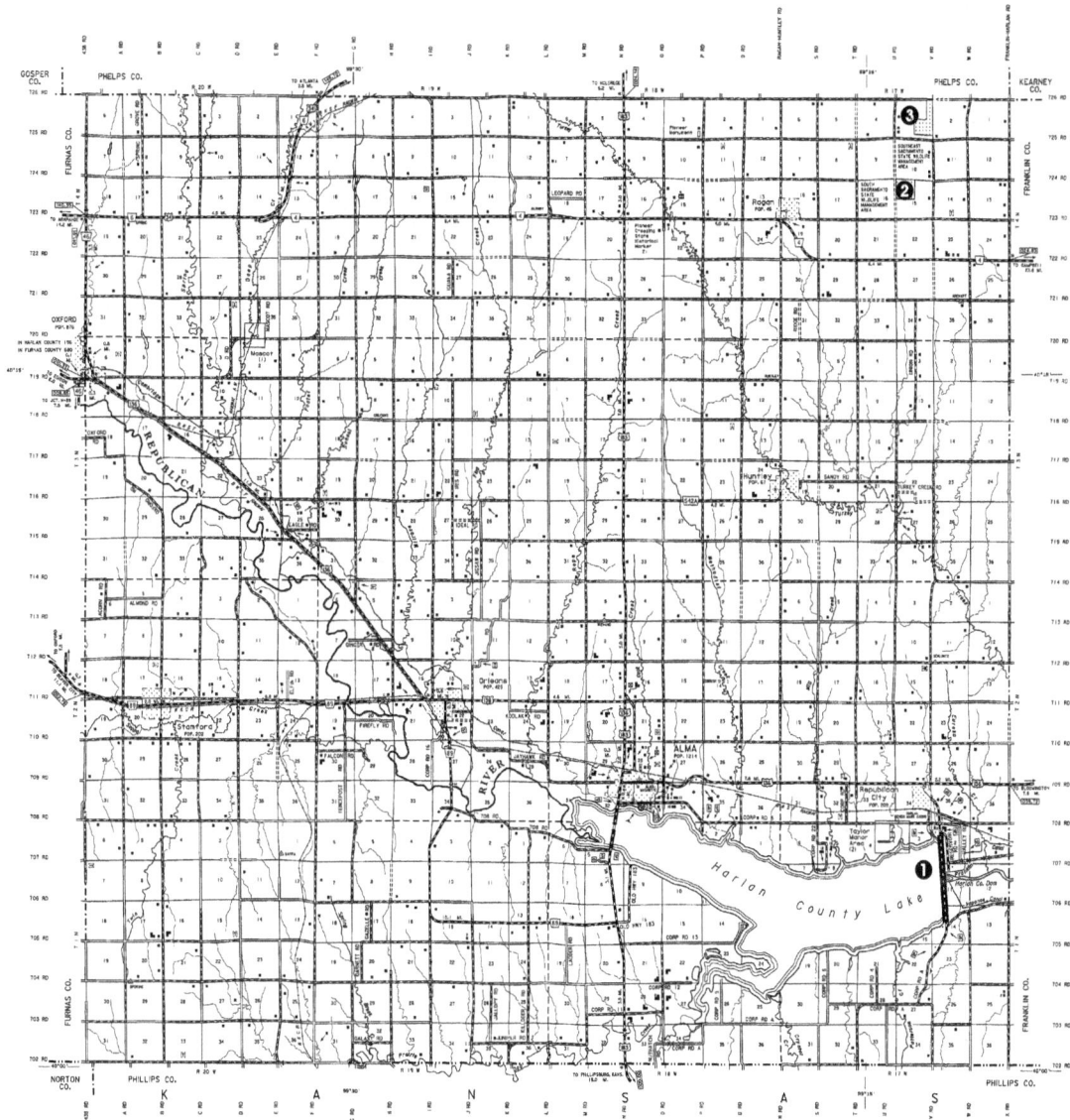

Harlan County   Map 20

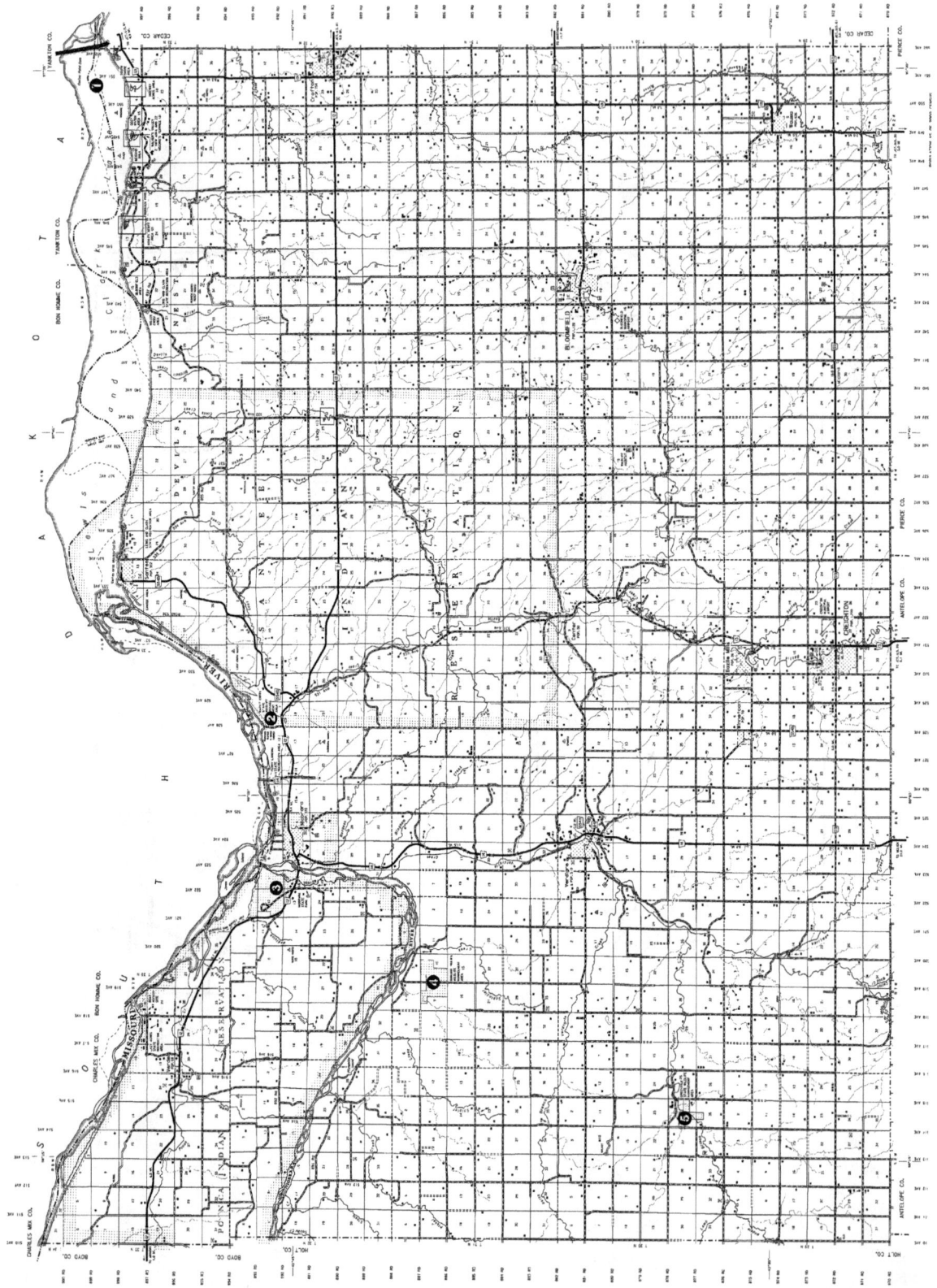

144

Knox County

Map 21

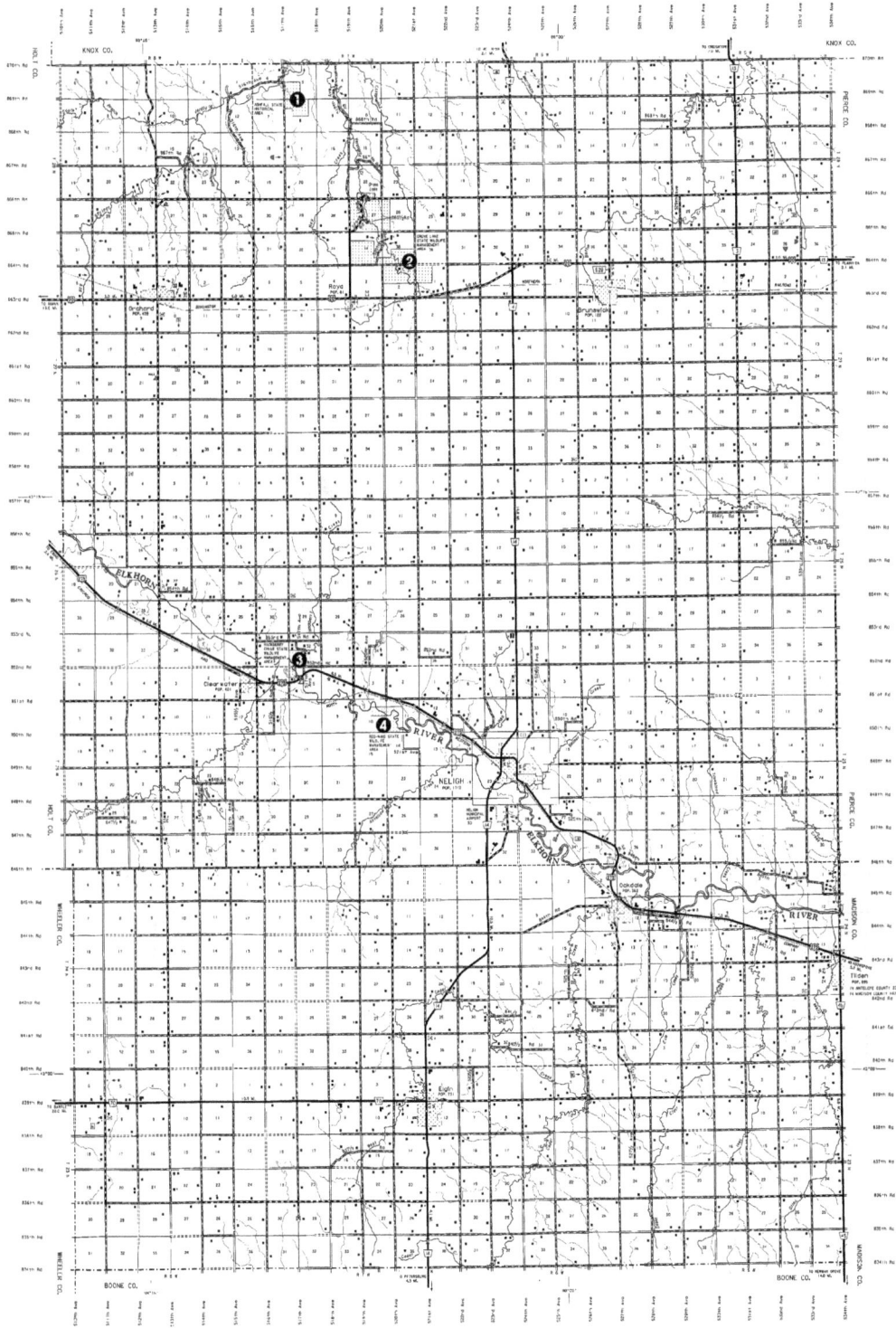

Antelope County   Map 22

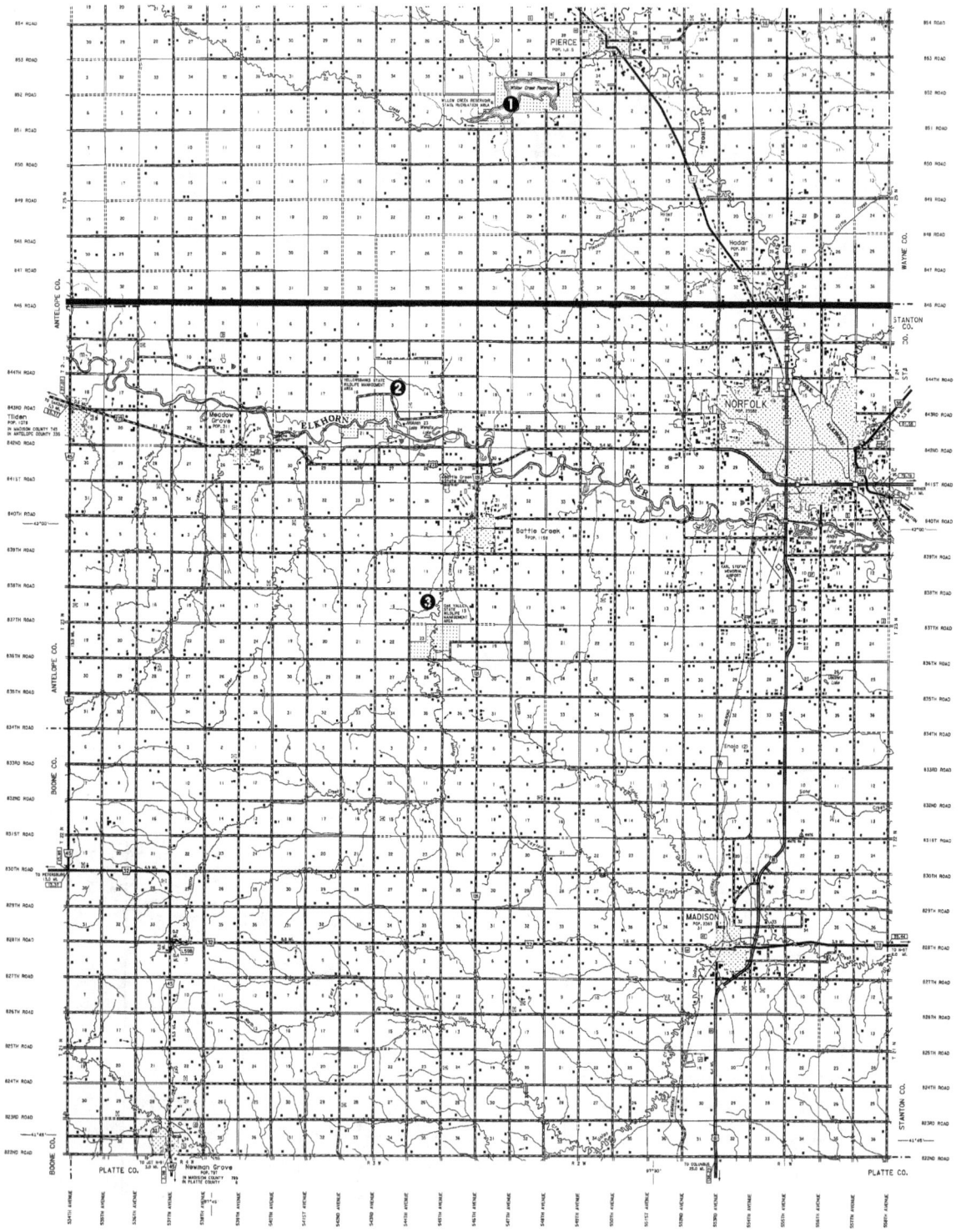

Madison and Pierce Counties   Map 23

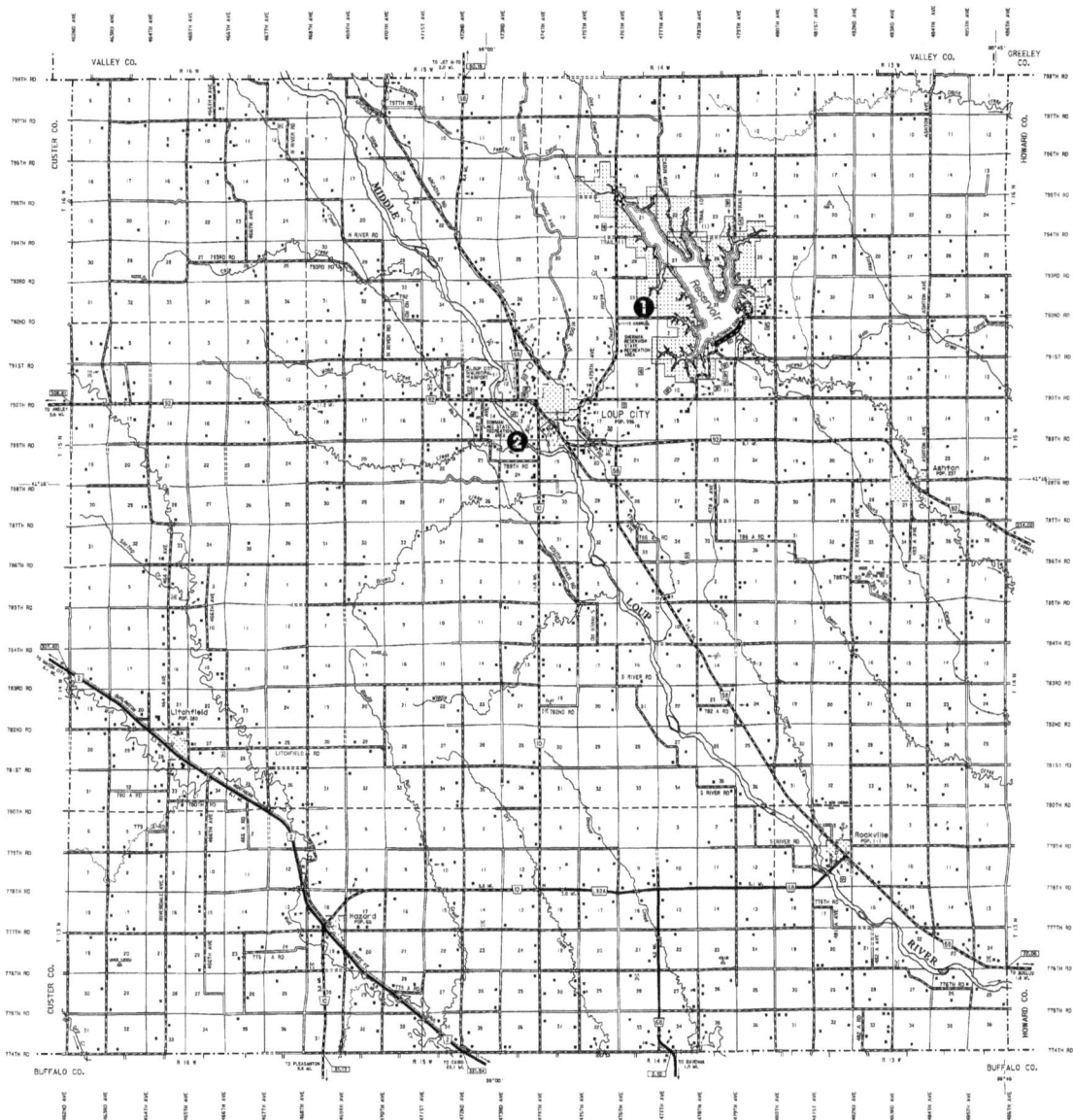

Sherman County   Map 24

148

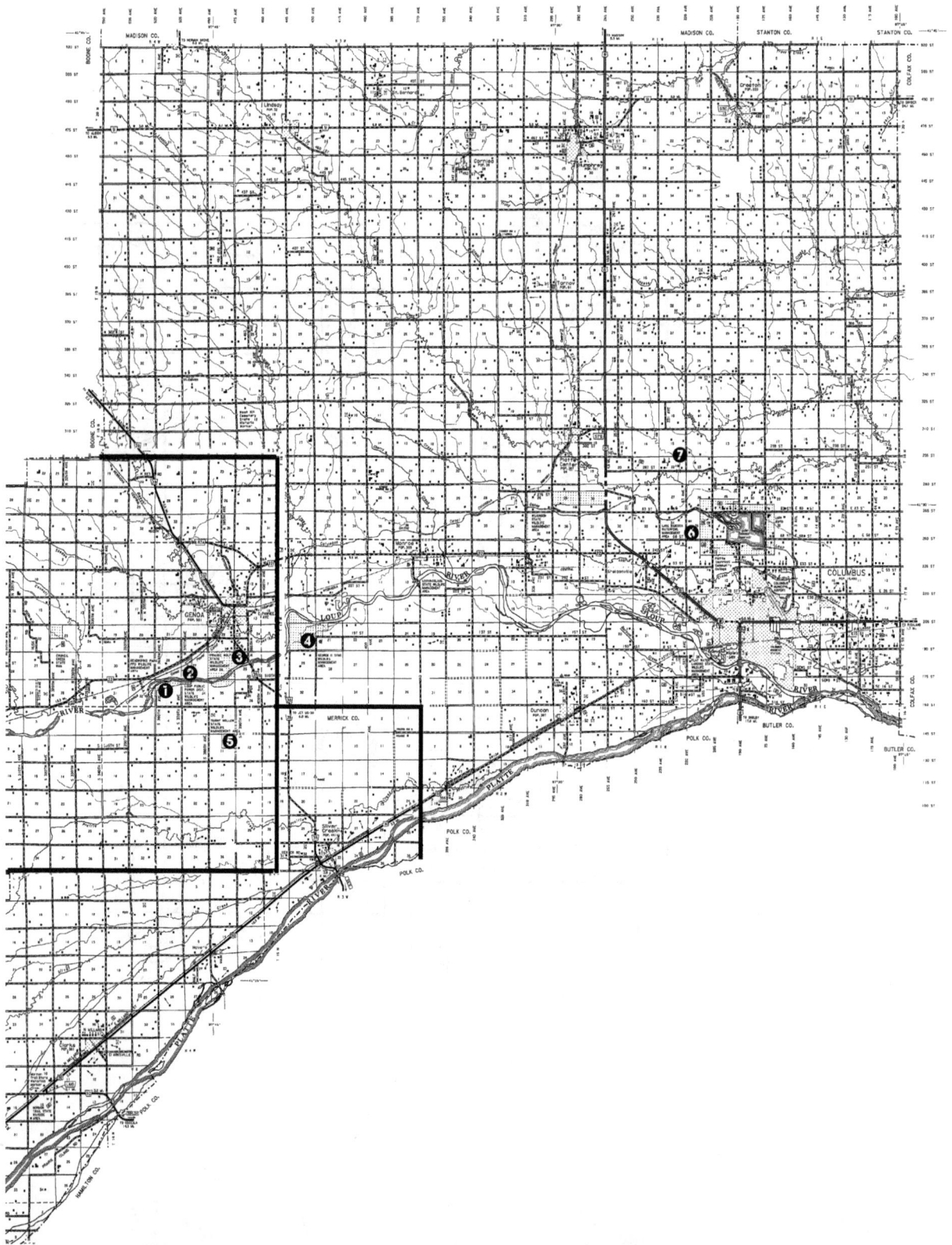

Platte, Nance and Merrick Counties

Map 25

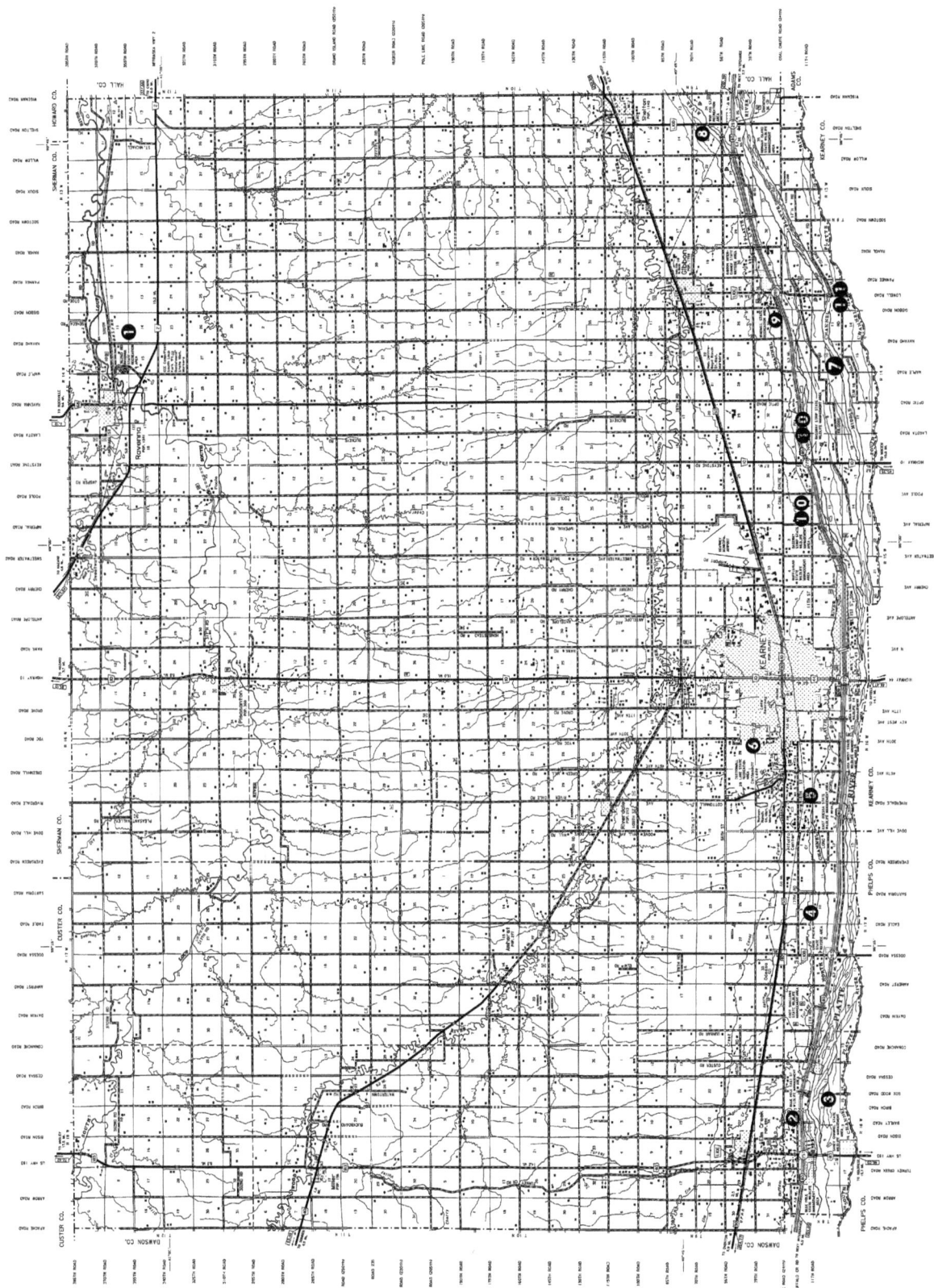

Buffalo County

Map 26

150

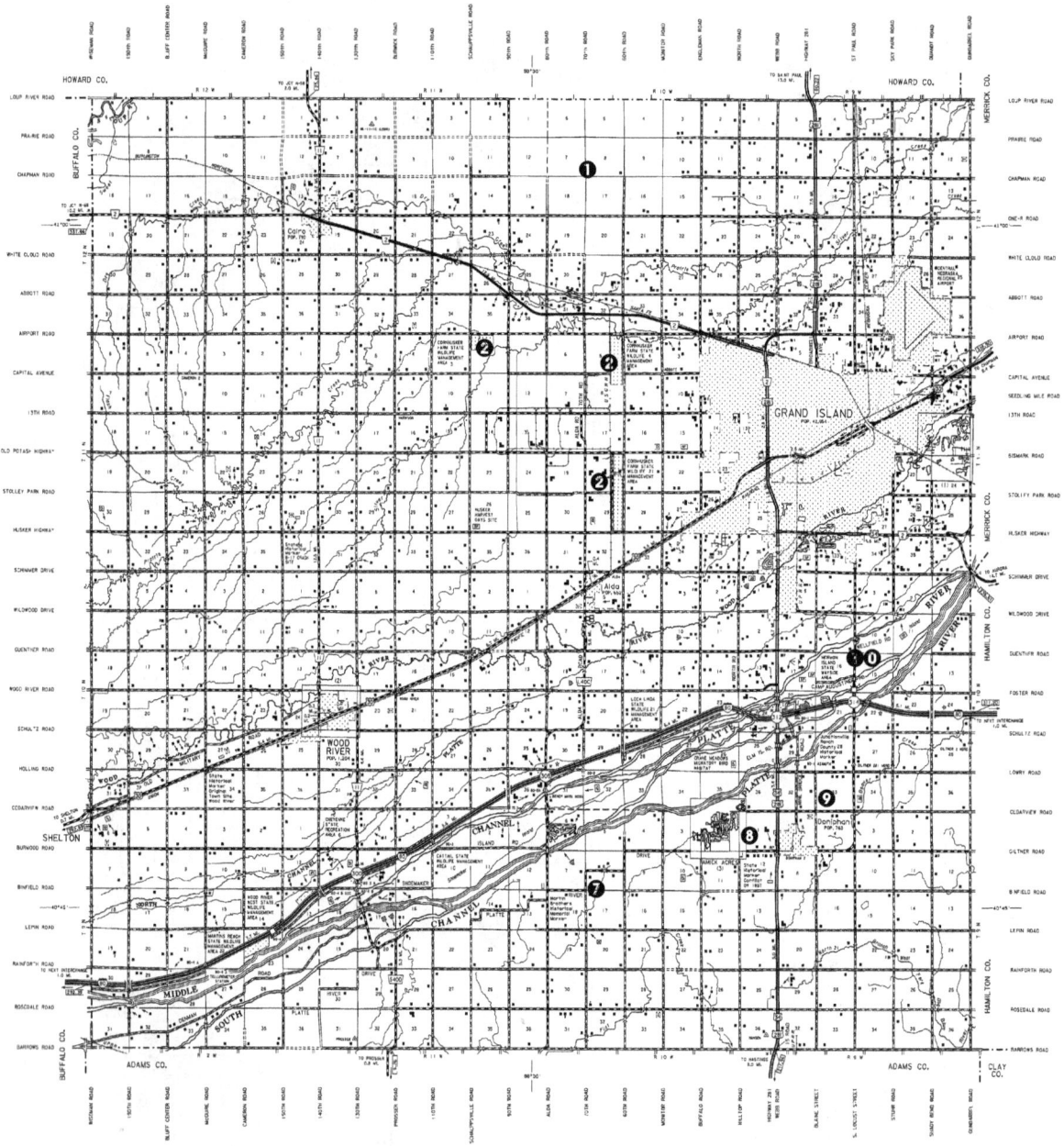

Hall County   Map 27

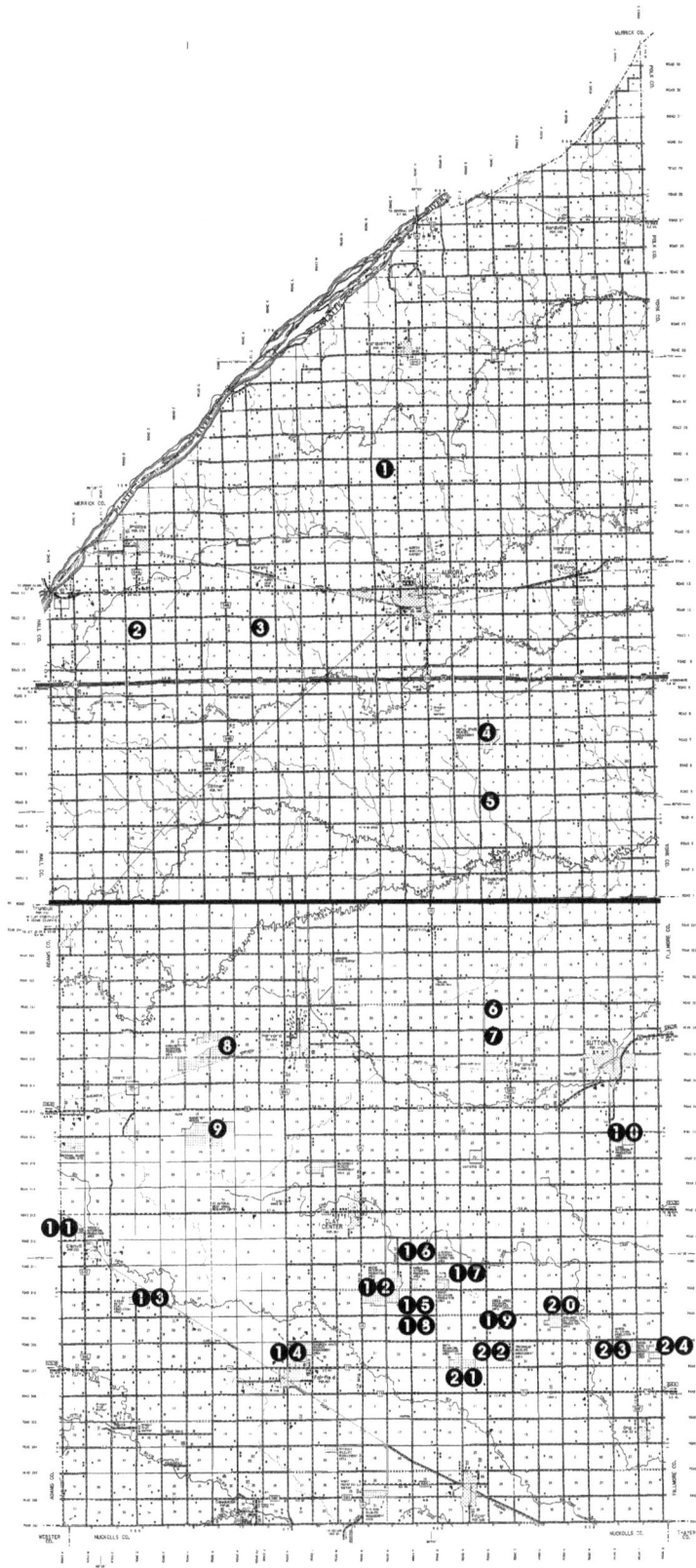

Clay and Hamilton Counties

Map 28

152

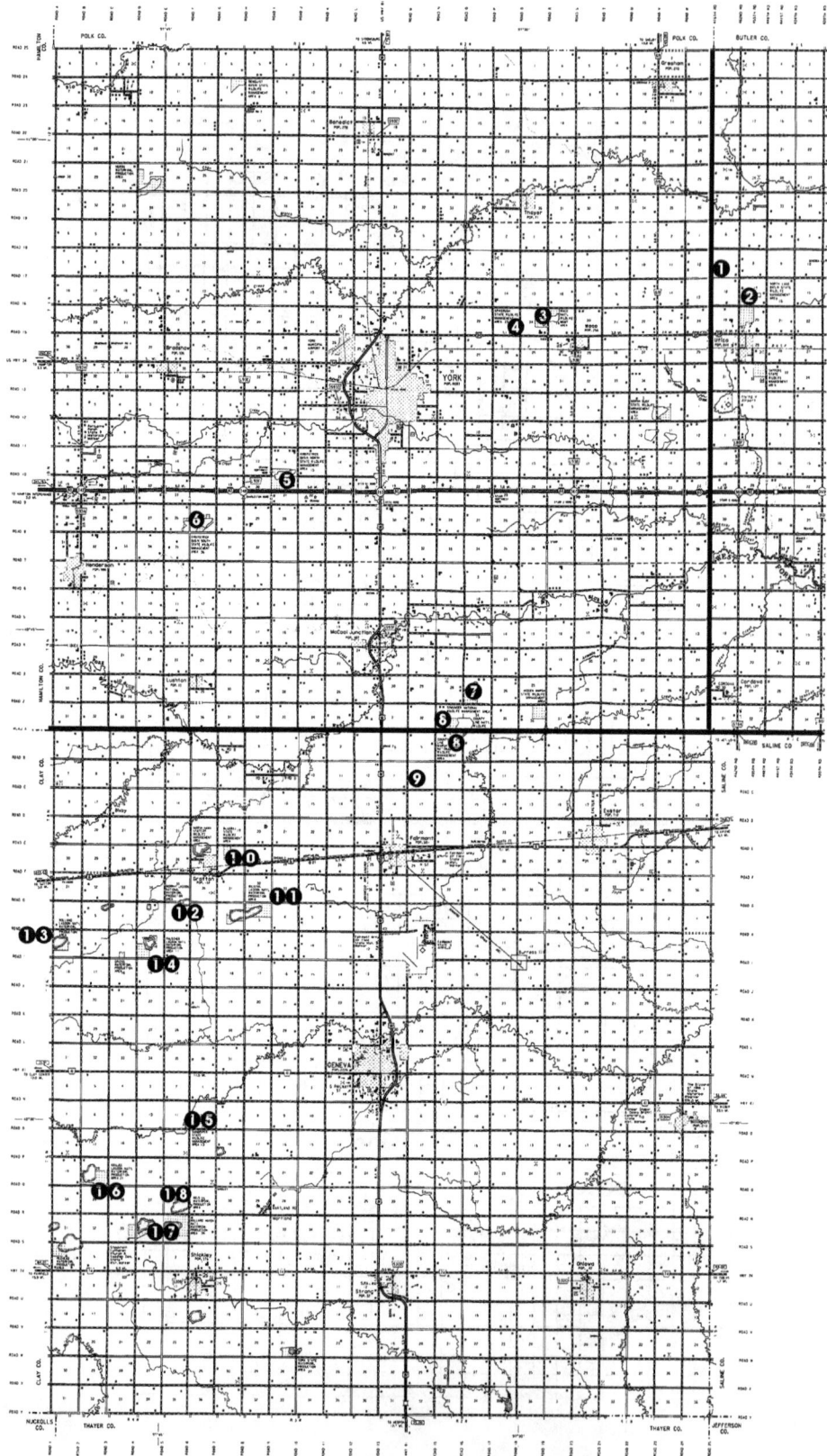

York, Fillmore, and Seward Counties

Map 29

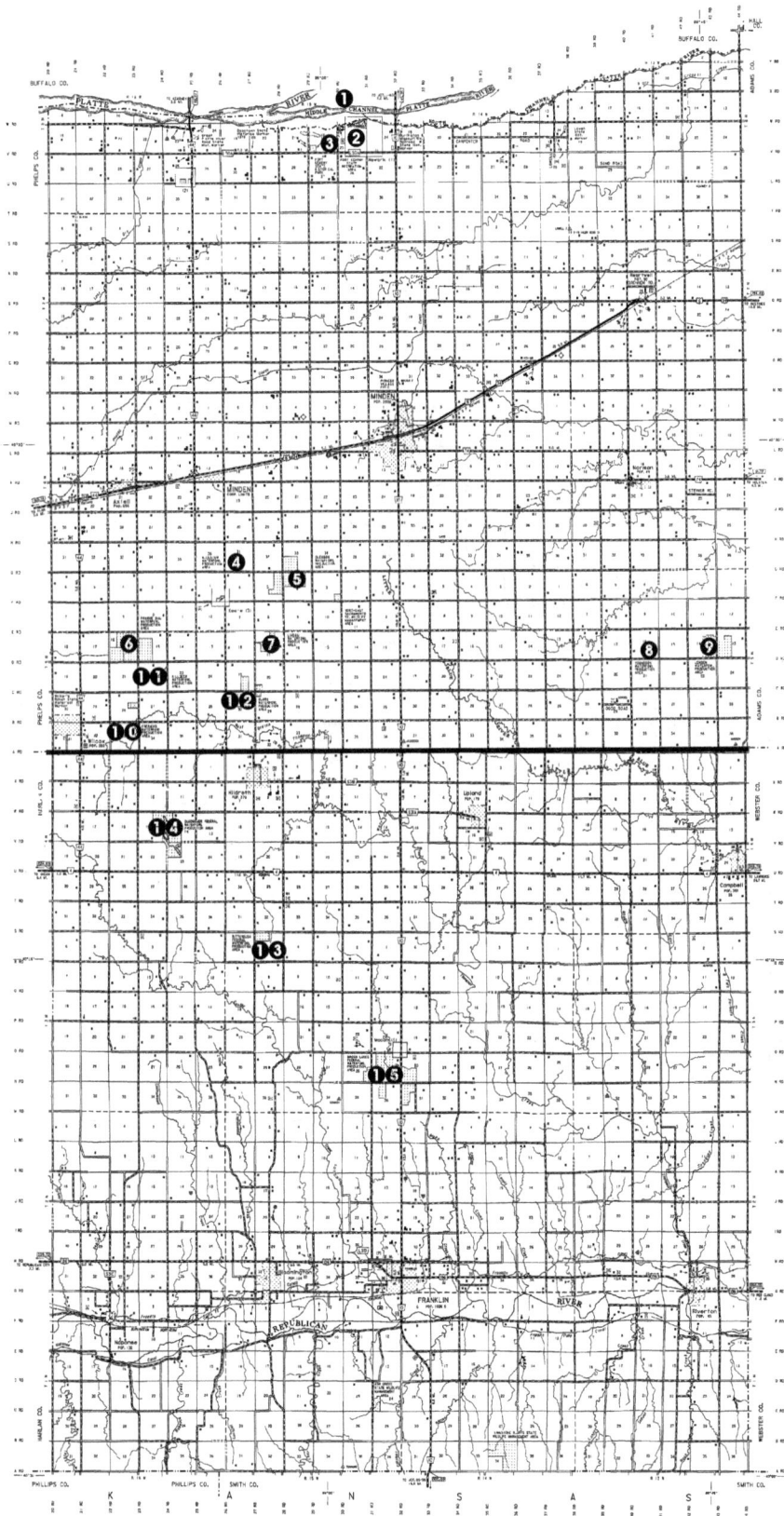

Kearney and Franklin Counties Map 30

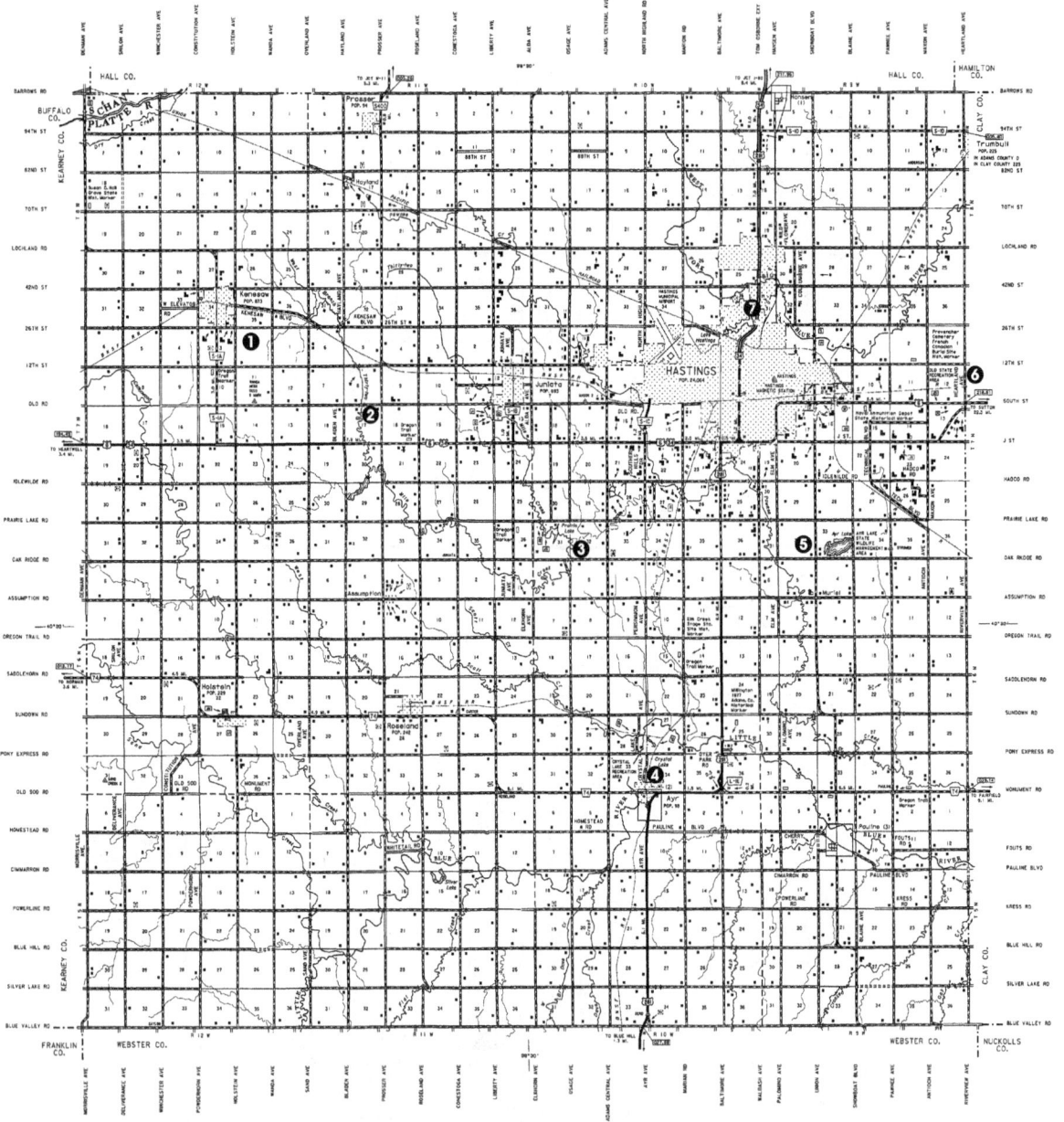

Adams County   Map 31

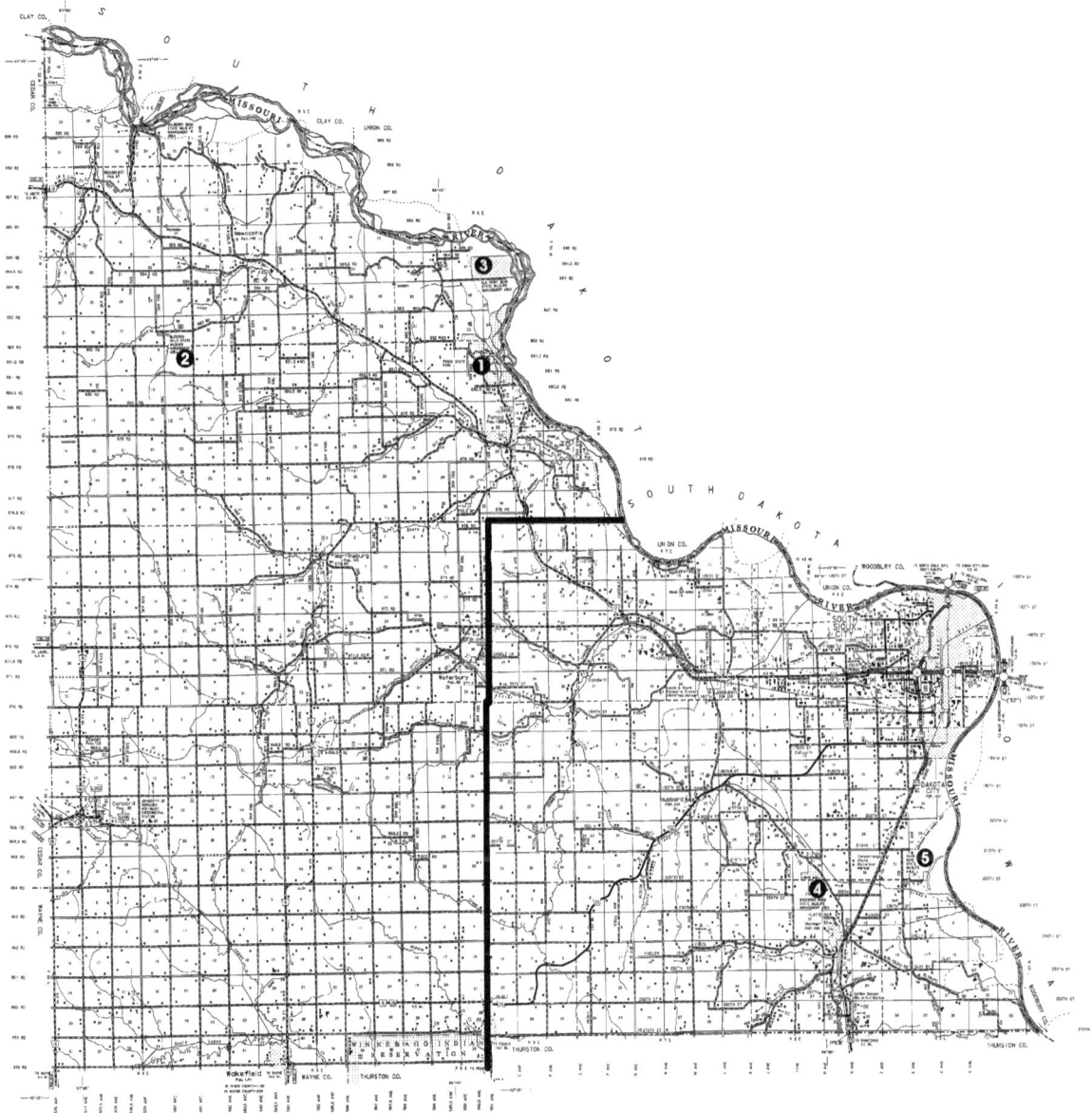

Dakota and Dixon Counties   Map 32

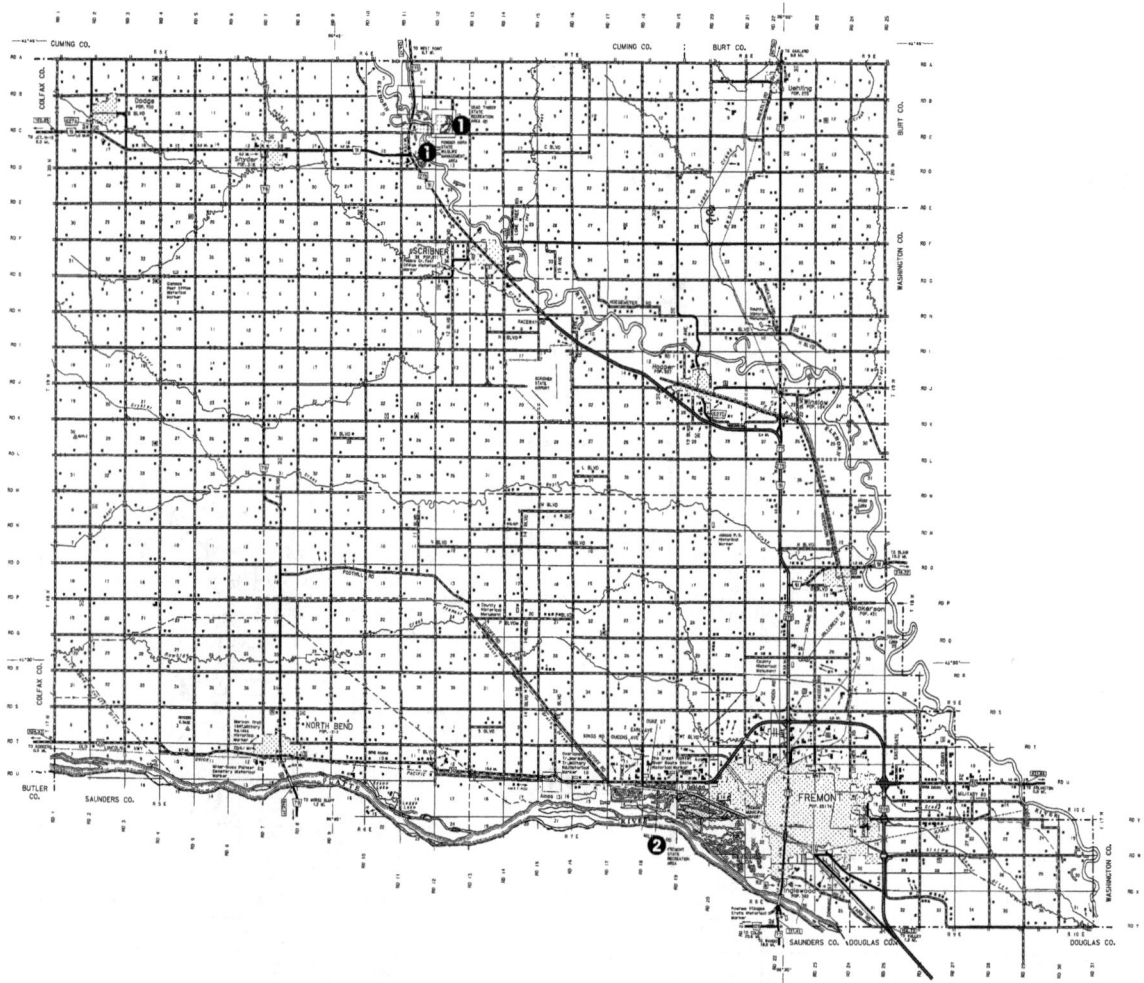

Dodge County   Map 33

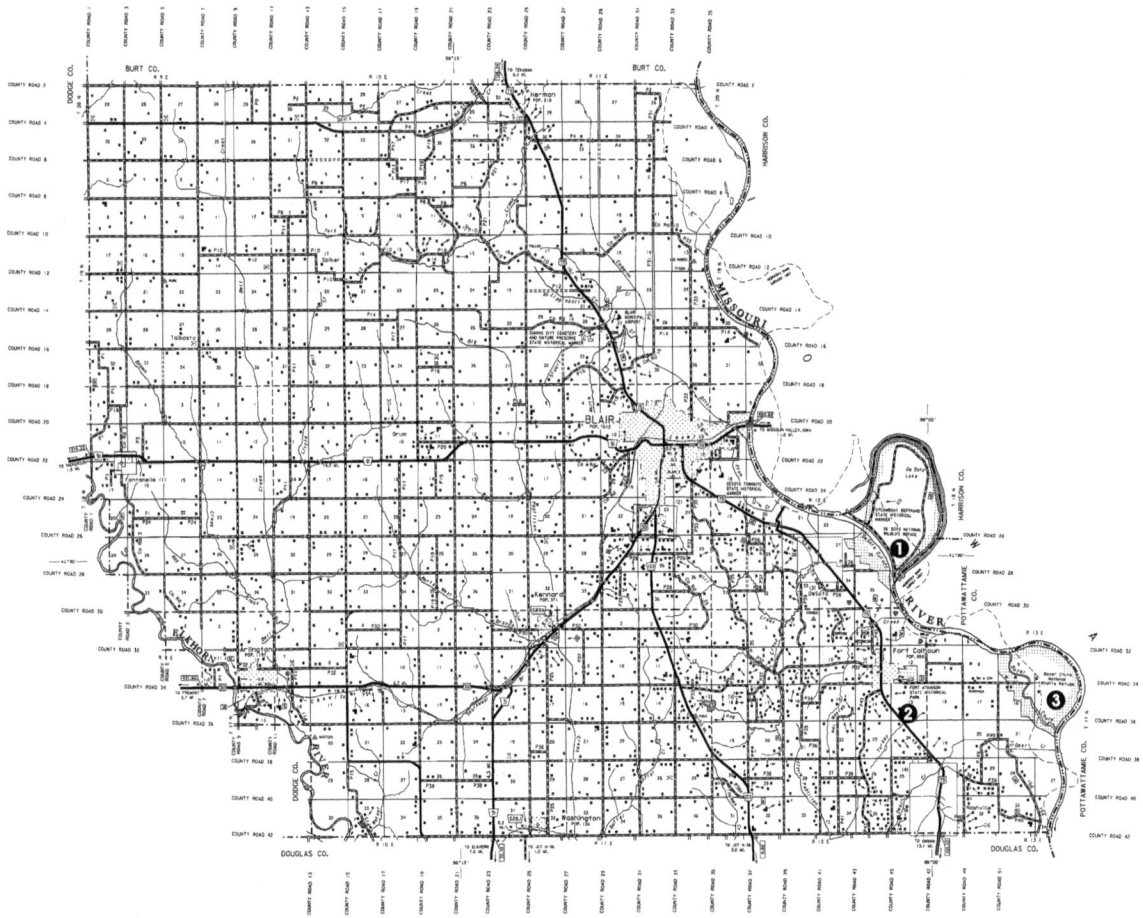

Washington County   Map 34

158

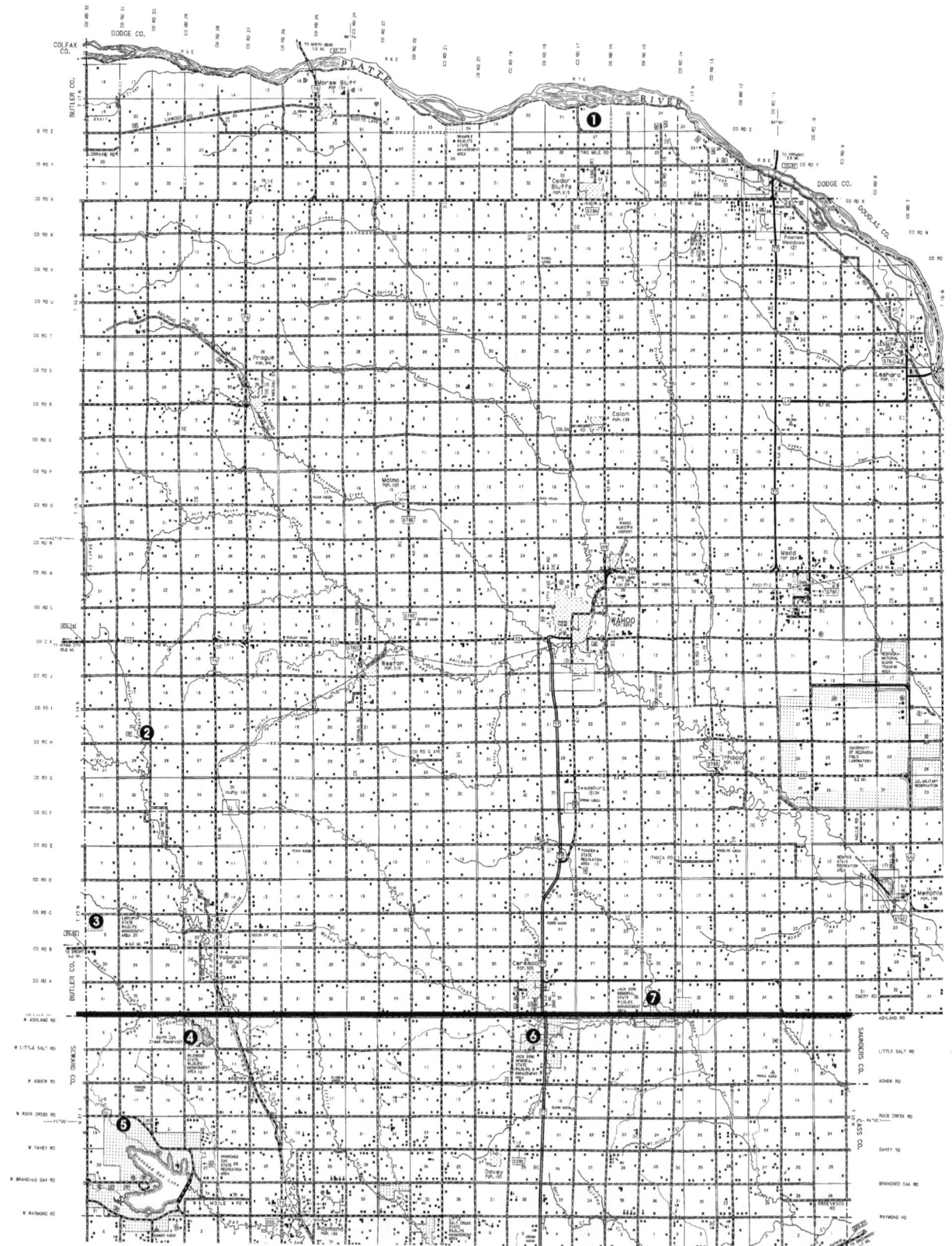

Saunders and Lancaster Counties   Map 35

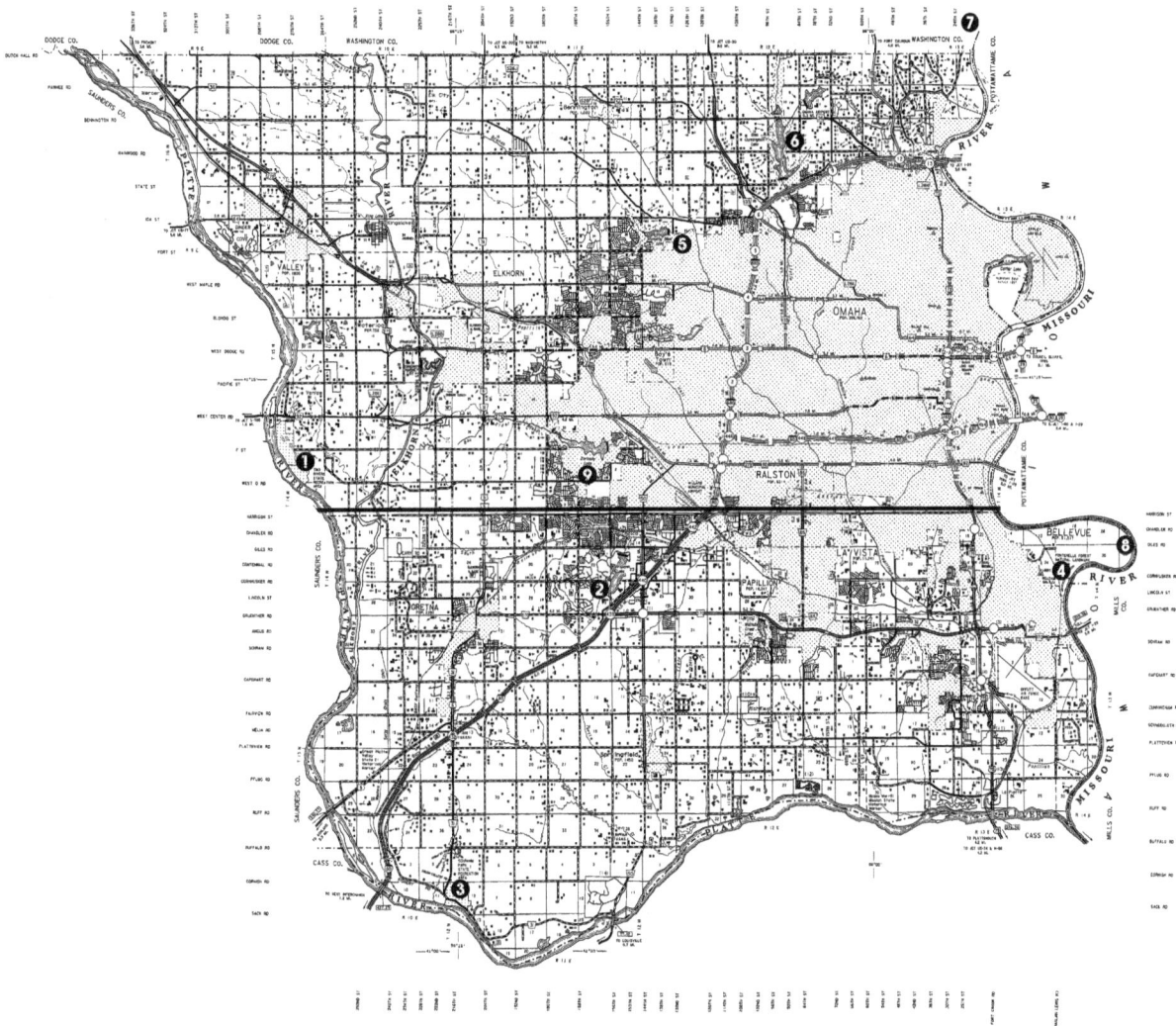

Douglas and Sarpy Counties   Map 36

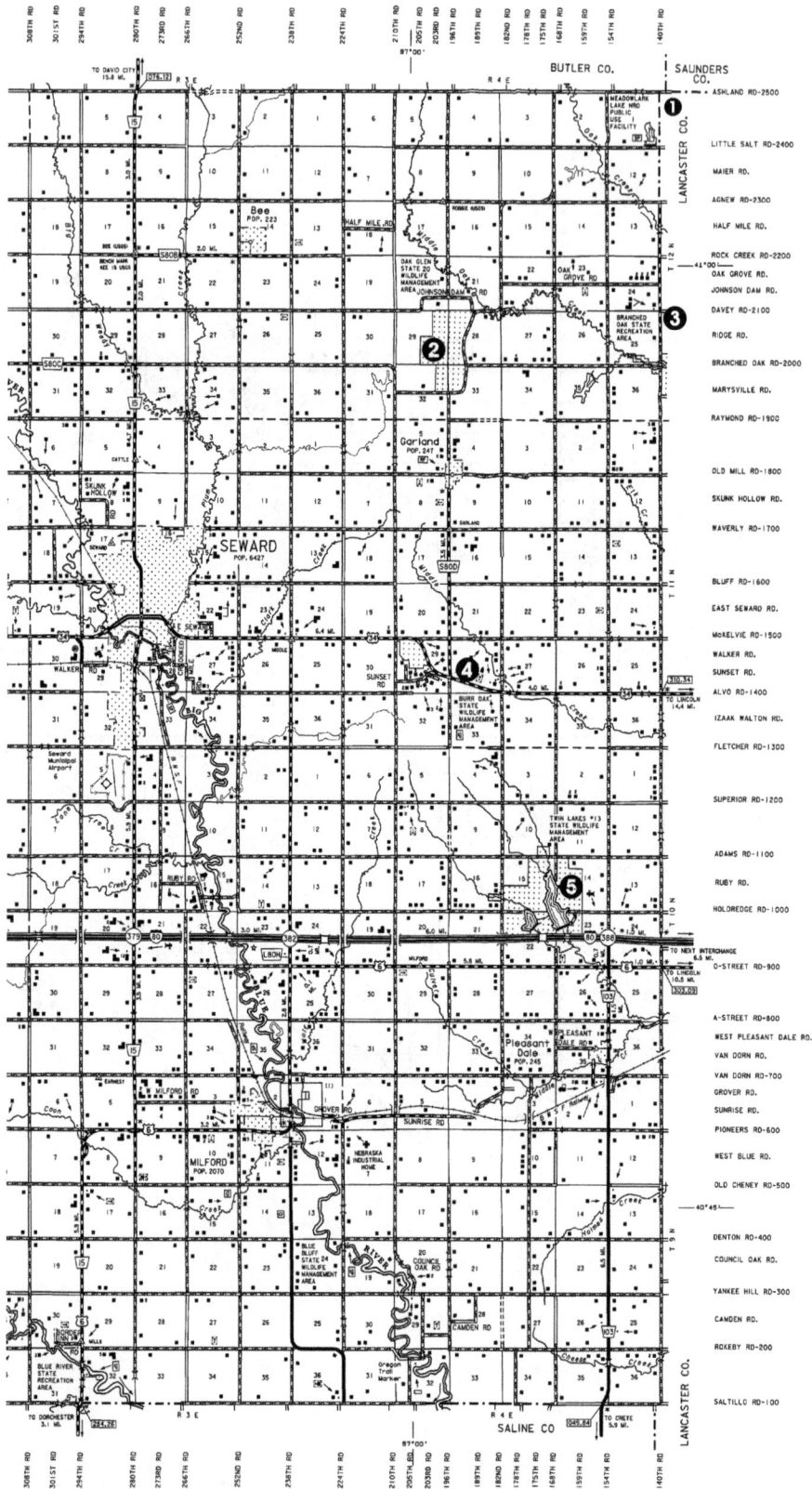

Seward County (Eastern portion)   Map 37

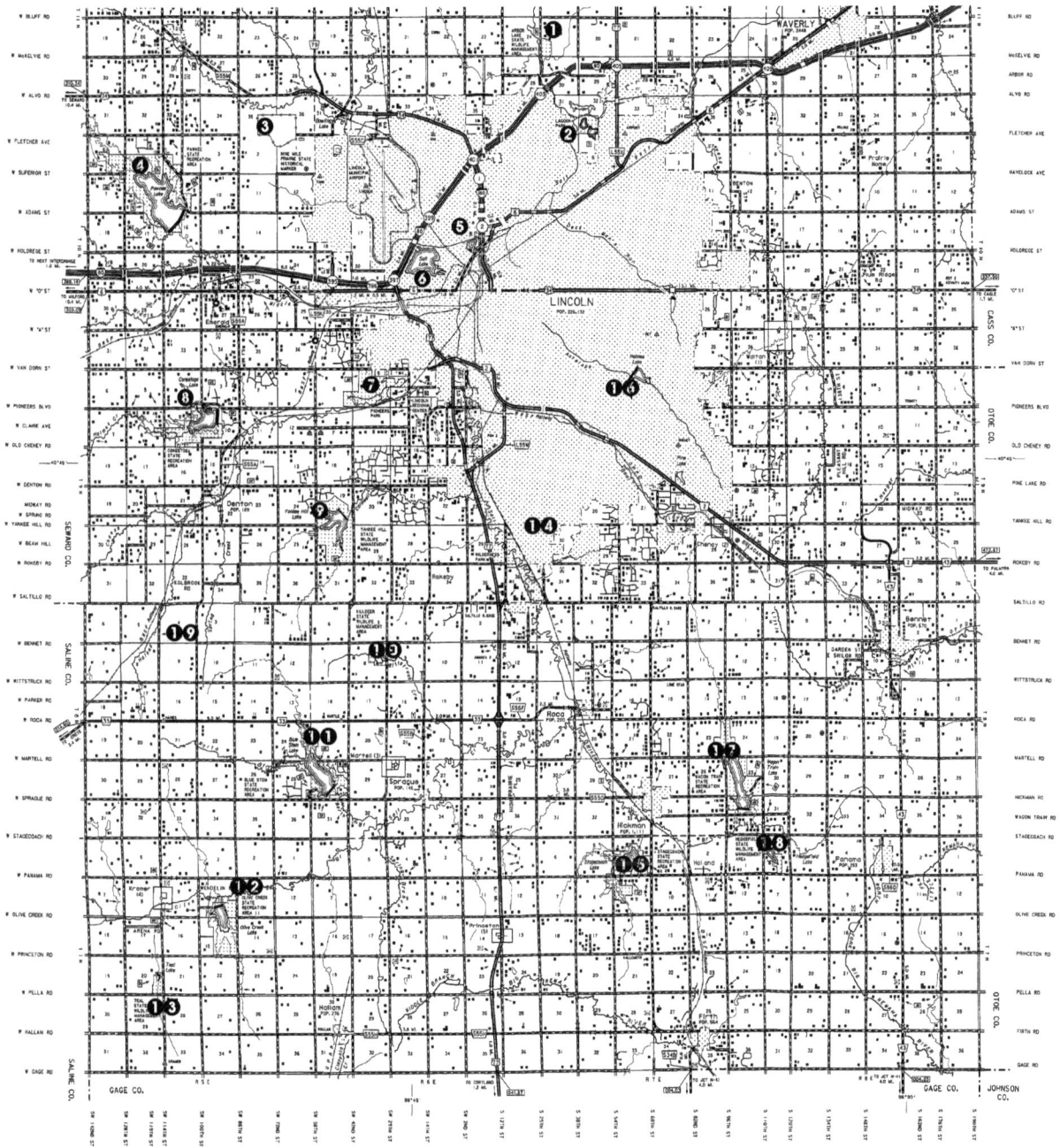

Lancaster County   Map 38

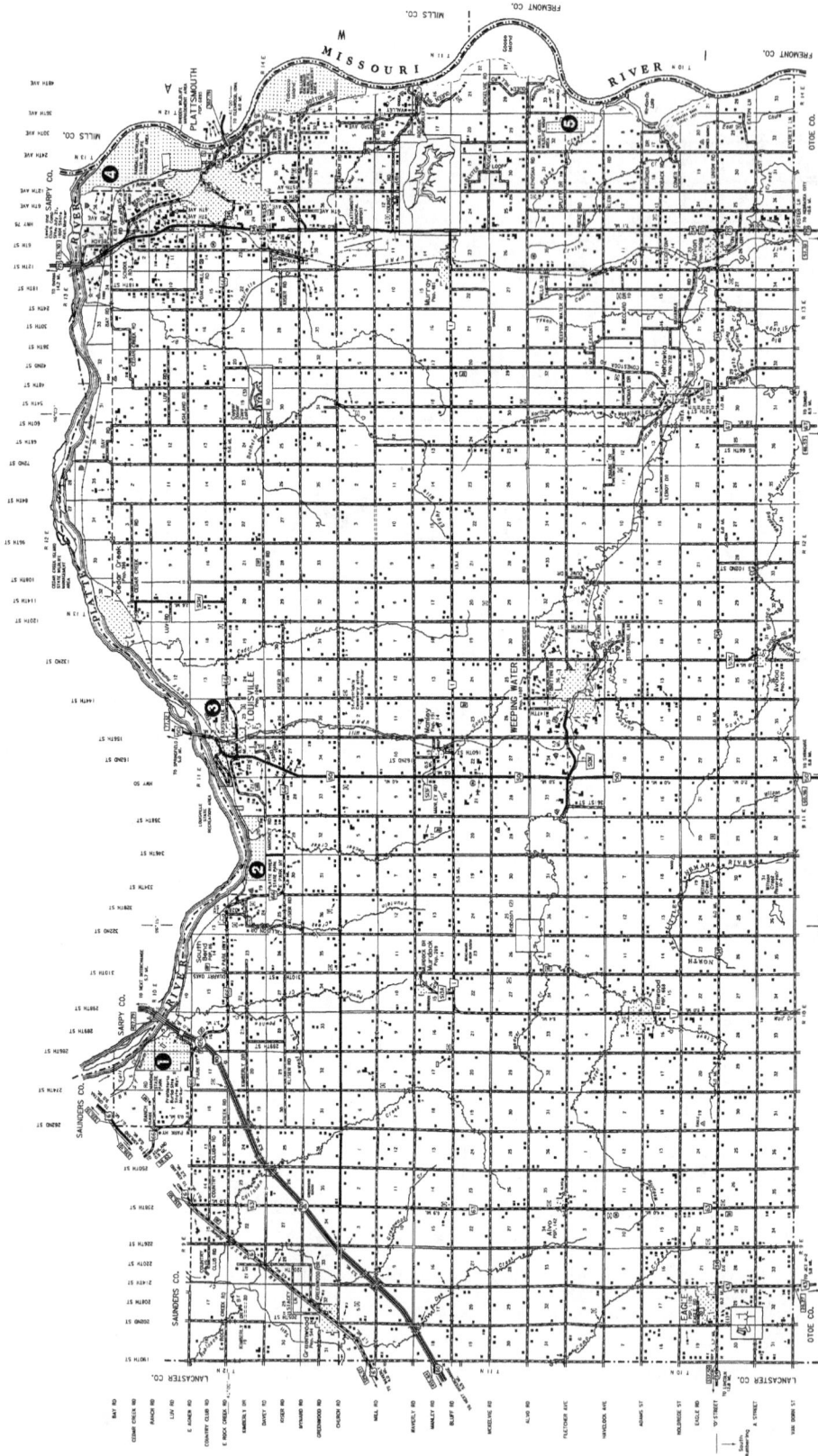

162

MISSOURI RIVER

PLATTSMOUTH

LOUISVILLE

WEEPING WATER

EAGLE

SARPY CO.

SAUNDERS CO.

LANCASTER CO.

OTOE CO.

MILLS CO.

FREMONT CO.

Cass County

Map 39

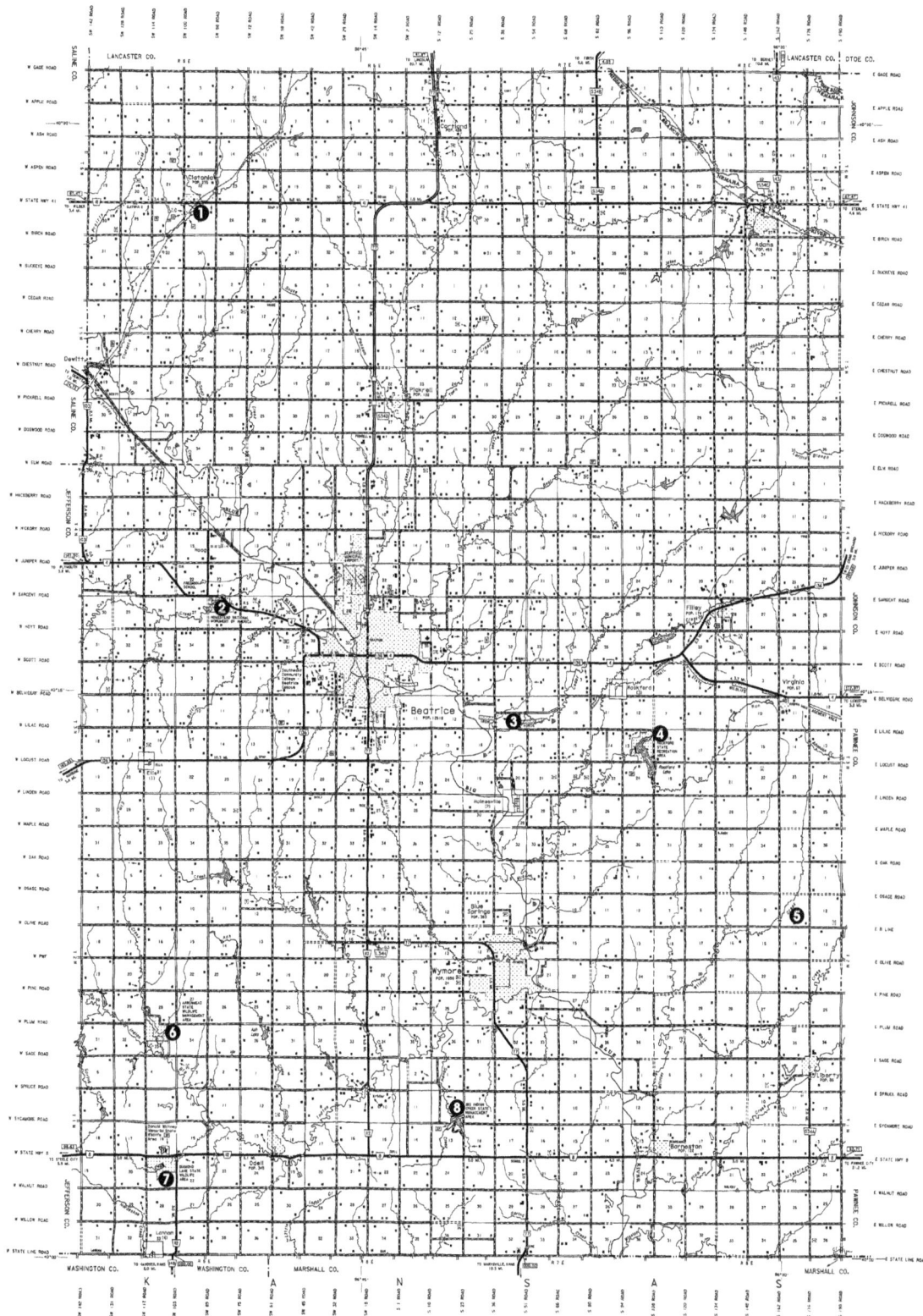

Gage County   Map 40

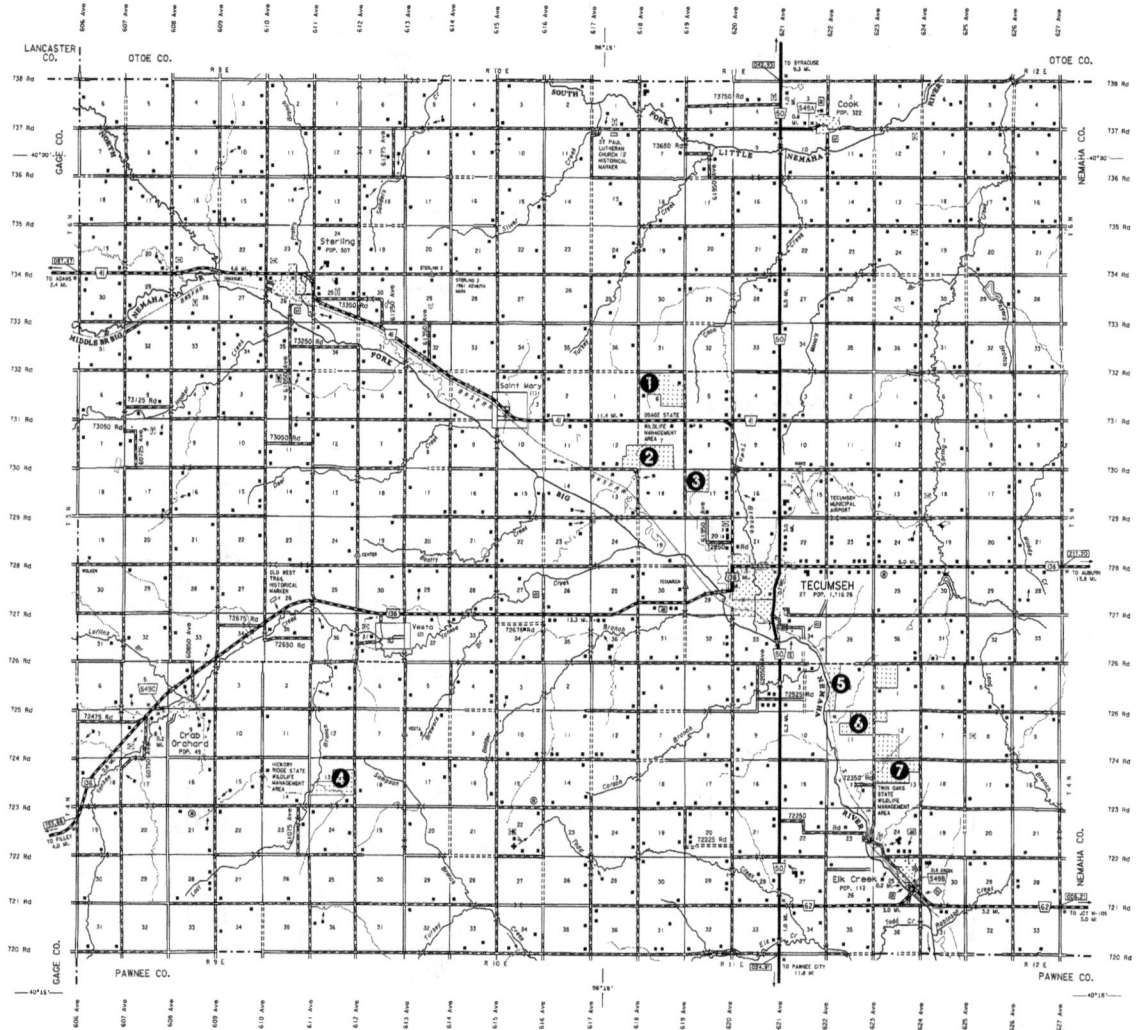

Johnson County

Map 41

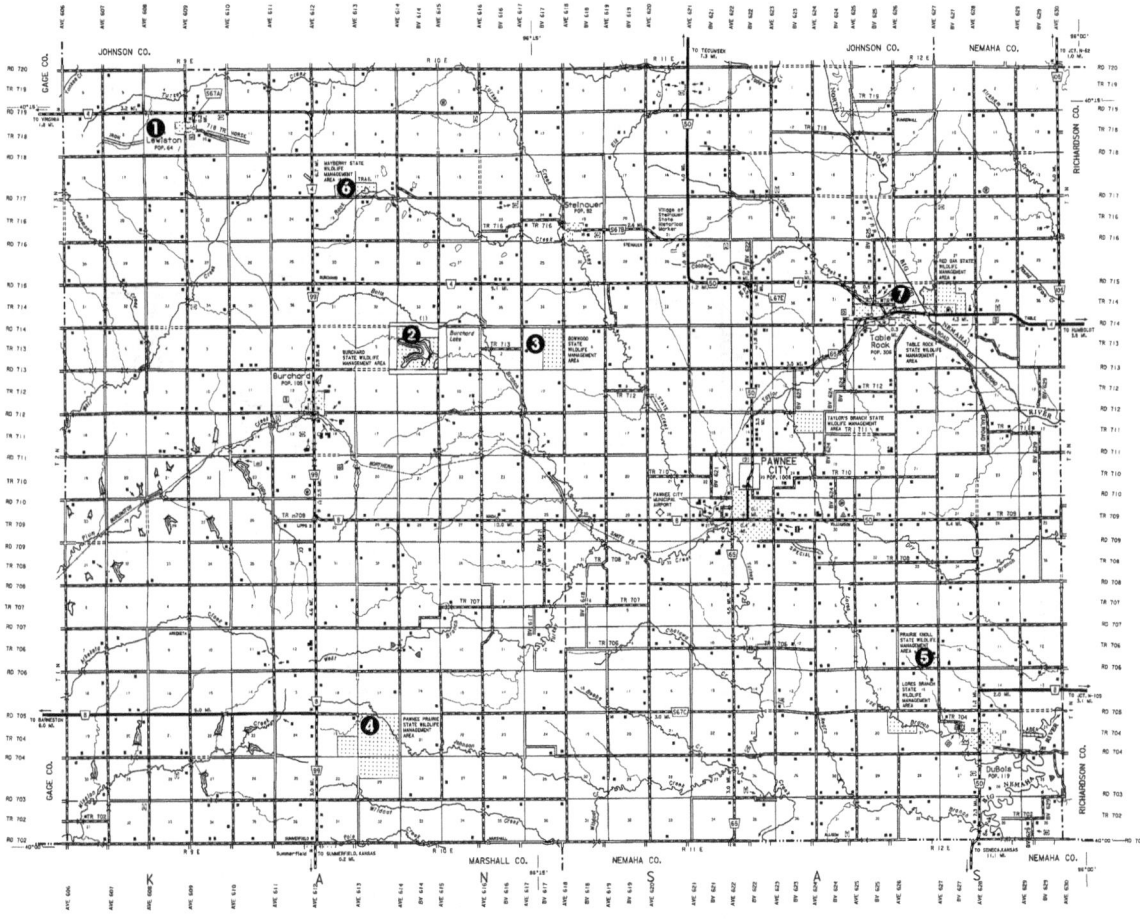

Pawnee County   Map 42

166

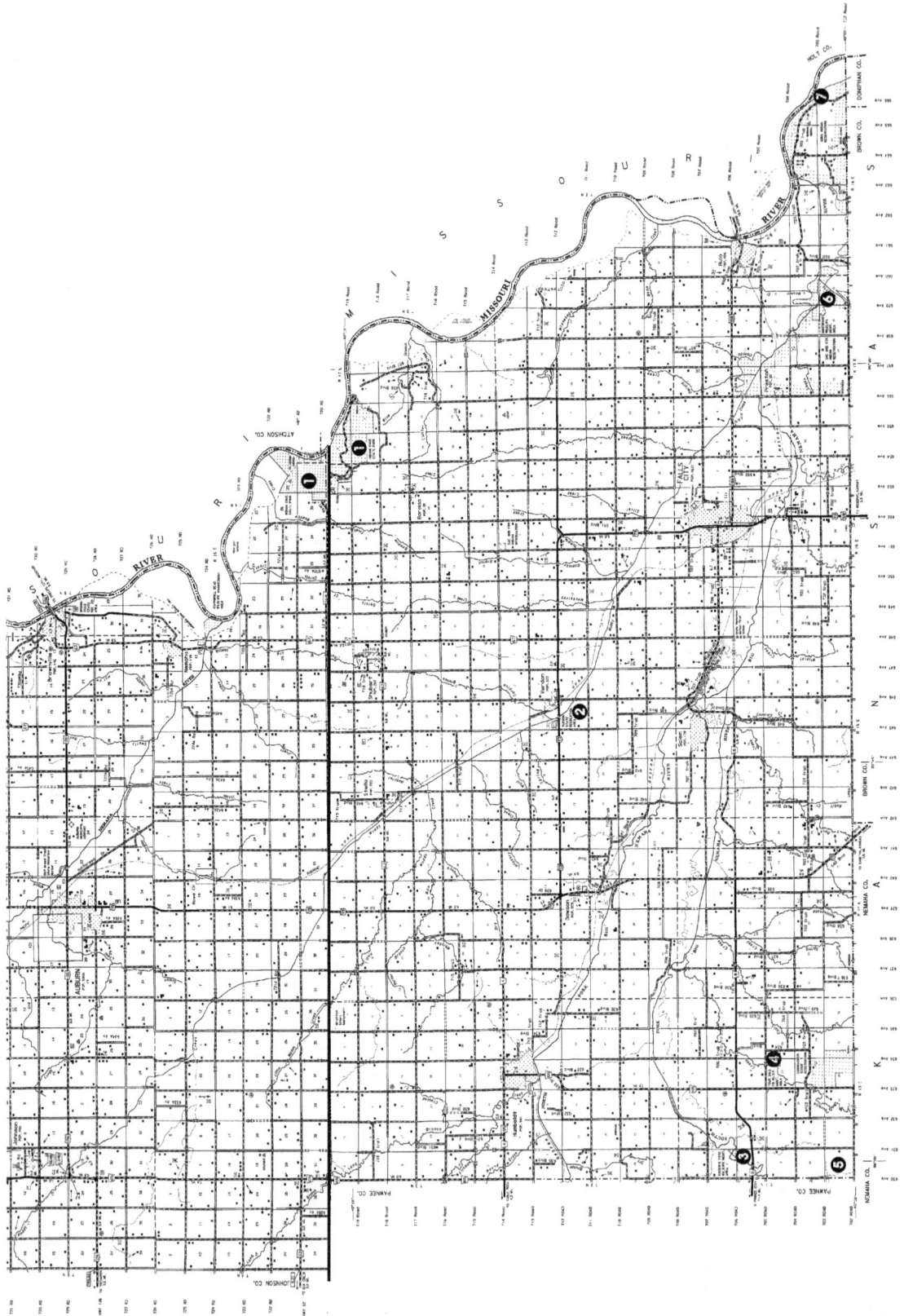

Nemaha and Richardson Counties

Map 43

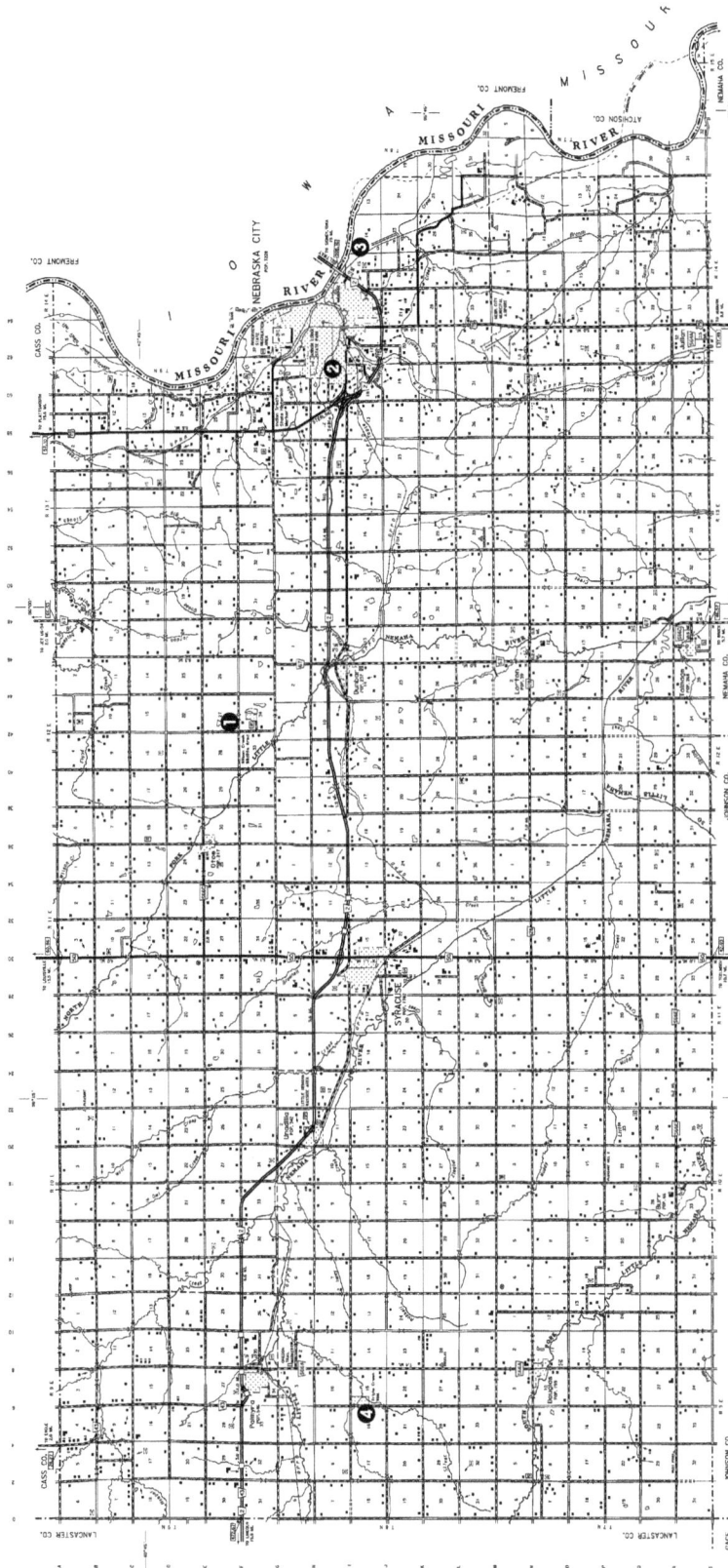

Otoe County

Map 44

9781609620110